SCIENCE ANXIETY AND THE CLASSROOM TEACHER

Donald C. Orlich

National Education Association
Washington, D.C.

Copyright © 1980
National Education Association of the United States

Stock No. 1679-3-00

Note

The opinions expressed in this publication should not be construed as representing the policy or position of the National Education Association. Materials published as part of the Analysis and Action Series are intended to be discussion documents for teachers who are concerned with specialized interests of the profession.

Library of Congress Cataloging in Publication Data

Orlich, Donald C
   Science anxiety and the classroom teacher.

   (Analysis and action series)
   Bibliography: p. 64
   1. Science — Study and teaching (Elementary) —
Psychological aspects. I. Title. II. Series.
LB1585.072   372.3'5'044   79-26926
ISBN 0-8106-1679-3

Contents

CHAPTER 1
FEARING THE UNKNOWN 5
Why Teach Science in the Elementary Grades? — Scientific Processes — The Characteristics of Inquiry — Questioning and Inquiry — Using Small Groups —

CHAPTER 2
MANAGING MATERIALS AND ORGANIZING FOR INQUIRY ..17
Using the Class Members as Helpers — Preparing for a Science Lesson — Other Coping Skills — Understanding More About Inquiry — Using Unguided Inquiry — Problem Solving as Inquiry — Evaluating Student Efforts — In Conclusion —

CHAPTER 3
USING QUESTIONS AND TEACHING SCIENCE 30
Formulating Meaningful Questions — Tips for the Teacher — Applying Three Strategies — Technical and Humane Considerations — Developing Student Skills in Framing Questions — Teacher Idiosyncrasies: A Caution — Summary —

CHAPTER 4
BEING SUCCESSFUL WITH SCIENCE-RELATED DISCUSSIONS..43
The Basic Organization — Why Use Discussions? — Introducing the Concept of Evaluation — Techniques for the Classroom — In General —

CHAPTER 5
FITTING THE PIECES ALL TOGETHER FOR SUCCESS........53
Organizing Support for Science In-Service Programs — Implications — Some Comments About Incentives — Teaching Students with Special Needs — Evaluating Science Programs — In Final Conclusion —

REFERENCES 64

RECOMMENDED READINGS 64

The Author

Donald C. Orlich is Professor of Education and Science Instruction at Washington State University, Pullman. He has also taught at Idaho State University and prior to that he was an elementary and junior high school science teacher in Butte, Montana. He is the author of *Designing Sensible Surveys* and co-author of *Teaching Strategies: Guide to Better Instruction*, *The Art of Writing Successful R&D Proposals*, and *Designs for Implementation*.

The Advisory Panel

Bernard W. Benson, Professor of Education, University of Tennessee at Chattanooga.

Wil Higuchi, K-12 Science Curriculum Chairman, Sidney High School, Sidney, Nebraska.

Kurt J. Johnson, Science teacher, Prince Georges County Public Schools, Maryland.

Lois Kenick, Science teacher, WLC High School, Wilton, New Hampshire.

Jane Butler Kahle, Associate Professor, Department of Biology and Education, Purdue University, West Lafayette, Indiana.

CHAPTER 1
FEARING THE UNKNOWN

During the 1960's and early 1970's our national concern for science reached, perhaps, an all-time peak. The advent of the Space Age made more people aware of the potential for scientific triumphs never before dreamed of. Science and scientists were accorded the highest esteem. Even at the time of this writing, our nation continues to place great faith both in our scientific and technological fields and in those persons involved in them. The mass media, especially television, have given the public a broad coverage of all fields of science and technology. As a result, these same media have contributed greatly to the stimulation of public interest in and understanding of science and engineering.

Yet, a comment frequently heard is that the schools of this nation have not focused adequate attention on science and scientific studies — this in spite of the fact that more and more scientific literature and supplementary learning materials have been made available to classroom teachers. The massive efforts of the National Science Foundation have stimulated elementary school science curriculum development to a state never before even contemplated.

In actuality, though, the "typical" elementary school teacher tends to be apprehensive about the teaching of science, despite the fact that science, technology, and scientific values permeate American culture. Perhaps that is the problem: Science is so much a part of our daily lives that we tend not to observe the evidence of its processes occurring within and around us, and, as a consequence, we think of *science* as a discipline exemplified by Albert Einstein's famous equations — understandable by just a handful of people in the world, and totally abstract to the rest of us. Elementary school teachers tend to ignore science because they have a high degree of anxiety about the topic, and, perhaps, fear that as teachers they do not know enough to teach it well.

But this need not be so. Science is similar to any of the other disciplines that are represented in the elementary school curriculum. Granted, the body of scientific knowledge is voluminous, and science has its own way of "knowing" — experimentation — along with a rather special vocabulary. But science is a means of generating new knowledge and, as one consequence, of making the world a better place in which to live.

The problems associated with teaching science at the elementary level have not resulted from indecision as to *what* to teach or even *how* to teach, but rather from what seems to be a *fear of teaching science*. To dispel this fear is my purpose here. My goal, and that of the National Education Association, is to help you, as an elementary school teacher, to better understand what science is and to become more confident in teaching it. I hope to provide a series of classroom-tested practices and techniques that will help each of you extend your knowledge of teaching per se to the teaching of one of the most fascinating fields known to humans — science.

Why Teach Science in the Elementary Grades?

There are many lists of objectives and rationales for teaching science to elementary school students. High on the list is the fact that science is having an impact on the lives of all people in the world greater than that of any other cultural facet, perhaps even including religion. You will note that scientific and technological knowledge is the factor used to distinguish among developing nations, industrialized nations, and "post-industrial" nations such as the United States. Thus, science impinges on people at their shops, homes, and transportation terminals, and even in their entertainment.

However, the materialistic well-being of a people is just one measure of the impact of science on the world's culture. Within our own nation we use science as one of the means by which to inculcate some of the following values:

1. Science is used to stimulate the critical and creative thinking skills of young people.
2. By knowing about scientific facts, principles, and theories, our citizens better understand and appreciate the whole of the planet earth and the universe.
3. The study of science aids in decision making by our government. Citizens with a better knowledge of the universe are better able to make more intelligent decisions — especially in the areas of the environment and those socially related considerations that are also a part of the world of science.
4. Scientific knowledge and its study lead to future careers in the field. These careers may not yet be invented. Recall that just a quarter of a century ago, there was virtually no computer industry, and jobs that required knowledge about electronic data processing were few. With just one tiny breakthrough — the transistor — the careers of millions of persons were created.

Among the other major goals of modern elementary science education, there is a genuine desire to stimulate the curiosity of all youngsters so that they can share in the excitement of scientific inquiry and investigations. Activity-oriented science programs allow students to conduct laboratory experiments that aid in the development of effective reasoning. These experiences also make young students familiar with the very methods and concepts used by real scientists in their daily work.

The Modern Rationales. The rationale of more contemporary science educators is that science is more than simple facts. Students must be given experiences that make them scientifically literate — i.e., make them aware of how a scientist works and of how the knowledge of science is generated.

A most important concern of the contemporary science movement is that students have direct experience with natural phenomena, that they be allowed to conduct scientifically related investigations. The use of activities means that the conceptual structure of elementary science is developed with appropriate teacher-provided guidance and experiences — points I will return to many times.

Finally, students learn that science and all the findings of science are to be construed as being divergent and not absolute. All scientific findings are quite tentative, and all facts subject to re-interpretation. New experiments and new findings continually change the meanings of previously known facts or concepts. The latter may be the most difficult concept for teachers and students at all levels to comprehend.

Scientific Processes

How does a student begin to comprehend the tentativeness of science? This comprehension comes from being immersed in the various *processes* usually associated with science. The Commission on Science Education of the American Association for the Advancement of Science has identified at least 13 processes that are critical to the learning of scientific reasoning. These processes are predicated on the assumption that scientists use a distinct set of intellectual processes.

Scientific literacy is developed by devising experiences that reinforce these processes. Through funding by the National Science Foundation, these processes were, in fact, implemented in a curriculum entitled "Science — A Process Approach" (SAPA). SAPA is unique in many ways. It is the first total curriculum ever to be constructed by using the processes of the scientific community. The processes are the content to be learned, while the scientific concepts and educational experiences are simply the vehicles through which the processes are applied. The writers and developers of SAPA identified generic scientific processes that are basic to all empirical endeavors. These processes are closely interrelated in the SAPA curriculum materials to show the relationships and the sequencing of the processes. Arranged somewhat in ascending order of complexity, the processes of SAPA follow:

1. *Observing.* Beginning with identifying objects and object properties, this sequence proceeds to the identifying of changes in various physical systems, the making of controlled observations, and the ordering of a series of observations.
2. *Classifying.* Development begins with simple classifications of various physical and biological systems, and progresses through multi-stage classifications, including their coding and tabulation.
3. *Inferring.* Initially, the idea is developed that inferences differ from observations. As development proceeds, inferences are constructed for observations of physical and biological phenomena, and situations are constructed to test inferences drawn from hypotheses.
4. *Using numbers.* This sequence begins with identifying sets and their members, and progresses through ordering, counting, adding, multiplying, dividing, finding averages, using decimals, and working with powers of ten. Exercises in number using are introduced to support exercises in the other processes.
5. *Measuring.* Beginning with the identifying and ordering of lengths, the development of this process proceeds with the demonstration of rules for measurement of length, area, volume, weight, temperature, force, speed, and a number of derived measures applicable to specific physical and biological systems.
6. *Using space–time relationships.* This sequence begins with the identifying of shapes, movement, and direction. It continues with the learning of rules applicable to straight and curved paths, directions at an angle, changes in position, and determinations of linear and angular speeds.
7. *Communicating.* Development of this process begins with bar graph descriptions of simple phenomena and proceeds through descriptions of a variety of physical objects and systems, and changes in them, to construction of graphs and diagrams for results observed in experiments.
8. *Predicting.* To teach the process of prediction, the developmental sequence progresses from interpolation and extrapolation in

graphically presented data to the formulation of methods for testing predictions.
9. *Defining operationally.* Beginning with the distinction between definitions that are operational and those that are not, this developmental sequence proceeds to the point where students construct operational definitions to problems that are new to them.
10. *Formulating hypotheses.* At the start of this sequence, the student distinguishes hypotheses from inferences, observations, and predictions. Development is continued to the stage of constructing hypotheses and demonstrating tests of hypotheses.
11. *Interpreting data.* This process is introduced with descriptions of graphic data and inferences based upon them. The student then progresses sequentially through the following activities: constructing equations to represent data, relating data to statements of hypotheses, and making generalizations supported by experimental findings.
12. *Controlling variables.* The developmental sequence for this process begins with the identification of manipulated and responding variables (independent and dependent variables respectively) in a description or demonstration of an experiment. Development proceeds to the level at which the student, being given a problem, inference, or hypothesis, actually conducts an experiment, identifies the variables, and describes how the variables are controlled.
13. *Experimenting.* This is the capstone of the integrated processes. It is developed through a continuation of the sequence of processes needed to control variables. This process includes interpreting accounts of scientific experiments, as well as stating problems, constructing hypotheses, and conducting experimental procedures.

It is assumed that there is progressive intellectual development within each process category. As this development proceeds, it becomes increasingly interrelated to the corresponding development of other processes. Inferring, for example, requires prior development of observing, classifying, and measuring skills. The interrelated nature of the development is explicitly recognized in the kinds of activities undertaken in grades 4 through 6, sometimes referred to as *integrated processes*; these include controlling variables, defining operationally, formulating hypotheses, interpreting data, and, as an ultimate form of such integration, experimenting.

So much for the real-life science illustration. Let us now examine how these processes and other elements form the inquiry attributes of science.

The Characteristics of Inquiry[1]

Science teaching and inquiry tend to be synonymous behaviors. To foster the art of inquiry, teachers of science tend to provide very little actual information, but, rather, they become the askers of questions.

Guided Inquiry. The easiest way to introduce any elements of guided inquiry is through the use of actual experiences. Before beginning the lesson, arrange all of the necessary materials so that students are all given materials with which to have similar experiences. When using guided inductive inquiry, do not expect the students to arrive at meaningful generalizations unless the learning activities, classroom recitations or discussions, materials, and visual aids are all available to the learners. Perhaps these initial experiences could be carried out through some small groups, such as task groups.

In conducting a lesson, make extensive use of question-asking skills. Ask the students to comment about what they actually observe. As they provide responses, be certain to distinguish carefully between statements based on observations and those based on inferences; when an inference is stated, simply ask, "Is that an inference or an observation?" To get students in the habit of being systematic, and this applies to all grade levels, ask each one to write the observation and beside it the inferences. This method also aids you in checking the observations that are the bases for any inferences. Of course, the latter technique cannot be used in kindergarten, grade 1, or even grade 2. In those cases, you can do the writing — or create a pictorial display of student observations and inferences. Of course, prior lessons with appropriate activities would have established the processes of observing and inferring.

Note how the 13 processes previously examined are integrated. As the class progresses, prepare a simple poster or use the chalkboard to list the actual observations and the accompanying inferences. Each process is slowly and carefully built with many examples drawn from actual experiences.

Patterns that the students observe are stated by them as generalizations that apply whenever the pattern is repeated. Thus, the process of inductive reasoning is gradually developed. Try to conduct guided inductive inquiry exercises whenever any occasion arises. Use simple experiences: "What could cause this type of track in the snow?" and "Where have we seen this before?" and "If we had to generalize about the sizes of these leaves, what one sentence could we make?" are the kinds of questions that require the learner to do the generalizing. In this manner you will seldom state the original generalization. You will, of course, be the one who plans for them.

Inferences. An inference is an interpretation or evaluation that results from making observations of objects or events. When you assemble a series of objects and ask the students to make generalizations about them, they will probably provide inferences. However, an inference is different from a generalization in that a generalization explains or summarizes some *verifiable* element.

For example, if you give students a big box of buttons and ask them to arrange the buttons according to some classification scheme, they will arrive at several different classes of buttons. They will also observe some trait that is general to all buttons in the group — three holes, shanks, shankless, round, square. If you ask for possible uses of the different buttons, you then approach the inference-building stage. The students begin to speculate about uses. But you ask other questions, such as "How could we find out?" or "Is that button big enough for an overcoat?" By asking your students to provide evidence or examples of how to test their ideas, you will help them reach the inference-building stage.

The Time Involved. When you initially utilize any type of inquiry activity in your classes, you must plan to spend at least two or three times the amount of time on each lesson as you might normally expect to. Any inquiry activity takes much time to plan, initiate, and complete. This time is spent on in-depth analyses of the content and on completion of activities by the students. Further, the use of inquiry methods requires greater interaction between the learner and the materials; there is greater interaction between you and the students also. Then, too, there is a bit of risk involved for the students. This is one of those times when the textbook does not list all the ideas and then prepare a "handy-dandy" summary statement. You will find that most children and adults approach inquiry activities with a bit of caution — if not apprehension. But as the inductive inquiry activities become a part of the ongoing procedures of the class, learner apprehension diminishes.

Another caveat is required: When you use any inquiry method, you will not "cover" the same amount of material that you do in expository teaching. Why? Well, you are using more time to develop thinking processes and are reducing the time spent on memorization of fact or content. Now this is your decision to make. You can't simultaneously maximize thinking skills and content coverage. If you desire to build the so-called *higher-order thinking skills*, then you must reduce some of the *content* and substitute *processes* for it. Really, though, you are not sacrificing anything. You are providing important instruction and experiences that are a part of the function of understanding the structure (epistemology) of science. *You* set that priority.

Incorporating Inquiry. In the early grades or when conducting initial experiences, the final generalizations might involve an oral summary or a review about the concepts, a listing of the ideas being presented, and, finally, the learners' own ideas as to what constitutes a meaningful generalization. As the learners provide such input, you would use the questioning techniques of reviewing and promoting.

Your class members can test generalizations by applying each statement to different times, places, or situations, or to other objects or events. A major limitation on this testing will be the backgrounds and experiences of the learners.

Another technique that may be used with guided inductive inquiry is to record a series of events on cards or some other manipulative medium. The latter could be a partially completed series or a constructed set of apparatus. Then ask the learners to place the events in the "proper" sequence — without any reference to being right or wrong. This activity could lead to a reading assignment. When the sequence of events is completed, ask the students to observe the pattern of events and to state the pattern in just one sentence. The latter would be the generalization. Other students could be asked to test the generalization by examining it for exceptions.

The above are tested techniques that help students to think inductively, to infer generalizations. As you use a variety of experiences, you will soon observe that the focal point of the inquiry session becomes a common experience for the entire class. These models are adaptable to all levels of instruction. The students' efficiency will improve with practice — and so will the teacher's.

Guided inquiry has at least the following characteristics:

1. The learners progress from specific observations to inferences or generalizations.
2. The objective is to learn (reinforce) processes or to examine events or objects, and then arrive at appropriate generalizations.
3. As teacher, you control the elements — events, data, materials, or objects — and, as such, act as the class leader.
4. The students interact with the stuff of the lesson — events, data, materials, or objects — and attempt to structure a meaningful pattern based on the observations made by each individual.

5. The classroom is, in a greater sense, to be considered a learning laboratory.
6. There is *usually* a fixed number of generalizations that will be elicited from the learners.
7. You encourage each student to communicate generalizations to the class so that others may benefit from unique perceptions.
8. Spontaneous responses that border on the creative or novel will usually be suggested by the students. In such cases, you should encourage even further novel responses.

The eighth characteristic needs some elaboration. Even though you might desire to provide experiences that have a fixed number of possible responses, there is always room for the open-ended inquiry activity. Teachers have long used the technique of asking students to "List as many ideas as you can for. . . ." When both you and the class reach this point, you'll be sharing in the real stuff of science — discovery.

Discovery behaviors may take at least three forms in the elementary school classroom: (1) the student has for the very first time determined something that is unique to her or him; (2) the student has added something to a discussion about a problem that you had not previously known; (3) the student has synthesized some information in such a manner as to provide others with a unique interpretation — i.e., demonstrated creativity. Student discoveries make school a fun place in which to work.

Questioning and Inquiry

Because questioning, as the most critical aspect of the concept of inquiry, stresses the *search*, the *investigation*, to accomplish these concepts, you become a *question-asker*, not a *question-answerer*. Teachers who are masters of guided inductive inquiry will tell you that they spend their time interacting with the students, but provide very few answers.

What Kinds of Questions Should a Teacher Use? Through extensive experience, several "stems" or lead-in questions have been categorized for those of you who want to bring about a more inquiry-oriented class environment. Of course, what makes these stems so interesting is that while they are specially designed for use in science, they are applicable to social studies, language arts, or mathematics classes — to *any* class in which you want to stress the process of inquiry.

When you are conducting an experiment, collecting data, examining cause-and-effect relationships, or analyzing events, the following set of question stems is the most appropriate by which to challenge the student to think:

> What is happening?
> What has happened?
> What do you think will happen now?
> How did this happen?
> Why did this happen?
> What caused this to happen?
> What took place before this happened?
> Where have you seen something like this happen?
> How could we make this happen?
> How does this compare to what we saw (or did)?
> How can we do this more easily?
> How can you do this more quickly?

Note that the above examples are oriented to *dynamic situations*. These stems are probably best classified as *prompting* questions.

When you are examining rather *static* living or nonliving objects, then the stems below would be most useful:

> What kind of object is it?
> What is it called?
> Where is it found?
> What does it look like?
> Have you ever seen anything like it? Where? When?
> How is it like other things?
> How can you recognize or identify it?
> How did it get its name?
> What can you do with it?
> What is it made of?
> How was it made?
> What is its purpose?
> How does it work or operate?
> What other names does it have?
> How is it different from other things?

These "prompts" help the students to understand better all kinds of interrelationships — one of the desired goals in inquiry *teaching* per se.

It is essential to keep reinforcing inquiry processes whenever possible, by simply asking, "What do we observe here?" or by stating, "Tell me about it!" It can be that easy. In fact, the two sets of stem questions given above might even be typed on a small card and used as a "prompter." (In Chapter 3, the skills of questioning will be developed in greater detail.)

Guided inquiry is an essential step toward developing more independent learning. The technique is dependent on effective teacher questioning procedures. The main point is that you constantly remind yourself: I ask the question; the pupils do the thinking and responding.

Using Small Groups

Small-group learning units are most appropriate for elementary school science to increase teacher–student and student–student verbal interactions in the classroom. Further, by using small groups, you add flexibility to your instruction. Teachers who use small groups report that this technique helps students reflect a more responsible and independent mode of learning. It is important to note, however, that you must identify and sequence a systematic and planned set of procedures, experiences, and objectives. During the actual lesson, students become active participants in the class activities, but in groups of two, four, or six.

How do you start? To initiate small groups or discussion groups requires that you simply restructure the manner by which individualized work takes place. Because most modern elementary school science programs have an activity orientation, you can begin by selecting an activity and then organizing a division of efforts among the class members. Each group member then has some specific task to accomplish — e.g., pick up and return the science materials, make a table to record data, set up the apparatus, conduct one aspect of the activity.

When learning activities result in more divergent types of experiences, it is more appropriate to have groups of four accomplish some assigned science objective. Collecting data or preparing histograms or tables can be difficult tasks for some students. By using a small-group approach you can help selected students become more competent in cognitive, affective, or psychomotor skills.

Teachers have collectively identified at least 12 goals or purposes for using small-group or discussion techniques in teaching elementary school science:

1. Interest can be aroused at the beginning of a new science topic or the closing of one.
2. Small groups can identify problems or other issues to be studied, or they can suggest alternatives for pursuing a topic under consideration.
3. A small group can explore new ideas or ways to solve problems, covering either the entire problem-solving cycle or just one phase.

4. Discussions provide an opportunity to evaluate data, inferences, sources of information, and methods by which the data are generated.
5. Small groups allow students to demonstrate individual strengths.
6. Students often learn faster and better from each other.
7. Students are provided an opportunity to use the vocabulary of science in an appropriate context.
8. Cooperative work skills can be developed through practice in small-group discussions.
9. Skills in leadership, organization, interaction, research, and initiative can be learned and improved through discussion techniques.
10. Ideas become more meaningful and personal if a student experiences them. Too, flexibility in understanding other viewpoints may be improved.
11. Discussion can provide the students (and you) with opportunities for learning to accept and value other ethnic and/or cultural backgrounds.
12. In a small group or discussion situation everyone can participate and feel good about that contribution.

After reflecting on these 12 possible outcomes from the use of small groups or discussions, ask yourself this question: "How many topics in my current science program lend themselves to a discussion, based on those 12 potentials?" If you are uncertain, examine your science program or textbook. Undoubtedly you will identify several topics that can be easily incorporated into a series of small-group discussions. Remember: A discussion is used to accomplish an objective — either process or content.

Now ask yourself: "What kinds of sharing experiences do I want for my students?" You probably responded to this question by focusing on sharing different cultural experiences. You may have thought about the need for students to display and share their unique talents. No doubt you thought of the disadvantaged and handicapped students, and of their need to be in a supportive and sharing environment.

The two preceding questions are important because you may be teaching in mainstreamed classes. Also you may teach gifted and talented groups. And, of course, you will be faced with the challenge of providing a nonsexist, multiculturally oriented education for all your students. By using small-group discussions you will help your students to meet the challenge of growing up in what may be culturally, ethnically, physically, and emotionally different environs — the public school.

CHAPTER 2
MANAGING MATERIALS AND ORGANIZING FOR INQUIRY

It has been my personal experience in working with virtually hundreds of elementary school teachers and administrators that their greatest problem is the preparing, organizing, distributing, and collecting of materials needed for any activity-oriented science program. Even if the program that you use has a "kit," many tasks must be accomplished before it can be used by the students. Here are a few tested "tips of the profession" from teachers who have eliminated this anxiety.

Using the Class Members as Helpers

The critical element in materials management is to involve one student or a small group of students in every management activity. In one sense, this means making every science period similar to a birthday party, including the packages. The only difference between a party and science is that in your classroom, the packages will be those essential components from the science kit or, if you do not have an organized kit, the materials from the science center. (If you have no center, we will cover that problem, too.)

Recall from the previous chapter how you can easily establish small groups. Well, you accomplish that exact task as an initial experience when beginning to teach science. Identify several groups of four. Allow each group to select a group name — Vikings, Seahawks, Gophers, Stars — and then place each team name on a small wheel. Along the outside of a larger wheel write a series of activities that must be accomplished. Then each time you teach science, move the wheel one "notch" so that all students participate in the materials management aspects of the science program. Figure 1 illustrates such a device.

But, back to the first science lesson, assuming that a kit is in use. Assign some seat work to the class and then gather your first group together to open the boxes in which the science materials are packed. I have watched students from 5 to 14 years of age become very emotionally involved in this activity as every package is a mystery.

You give some general directions, such as "Joan and Suzie, take the big box down to the custodian for storage." In this manner, the packing crate or box is disposed of in a proper manner. Then give the remainder of

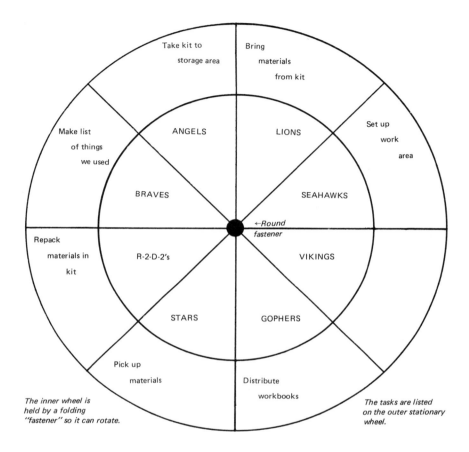

Figure 1. Science Classroom Task Organizer

the students specific instructions on how to inventory the items. You will find that each kit comes with some type of inventory checklist. By checking what is available immediately upon unpacking, you will know what materials need to be requisitioned from the office or delivered from the school district's central storage facility.

Make yourself familiar with the organization of the kit. If a "map" of the kit is neither included in the kit nor available through your central office curriculum staff or science consultant, have your students draw one for your use. The map should identify the trays and drawers, as well as the contents of each.

If your science program does not have a kit, then you and your students must become "explorers" in order to assemble the necessary equipment and supplies. The easiest way to handle the problem is to refer to the teacher's or student's guide. Make a list of the needed materials or equipment, locate them, give the list to a small group of students, and send them off to the storeroom or to wherever such materials are kept.

One way that you can collect materials beyond what must be provided by the school district to implement the science program is to send home a "scavenger hunt" announcement. List all the inexpensive items of equipment or supplies that people tend to have around their house or apartment. Again, the class will be the key to the success of the scavenger hunt. Schools that sponsor scavenger hunts often receive more materials than they can possibly use and can then share their good fortune with schools needing similar items.

When the materials have been gathered for the lesson, you must then develop a system that the students can use to organize the materials. Use shoe boxes, or any similarly sized boxes, for storage. Label each box as to its contents and, most importantly, the concept it is being used to teach. Also identify the text or program used, the related page numbers, or the activity for future reference. In your teacher's edition of the science program write down what materials you have, where they are stored, the quantities you have of each item, and the success you had with the activity. The latter information will prove helpful the next year that you teach the activity.

Of course, the students compile the inventories for the boxes. This little task teaches responsibility, a most important facet of life. It is important to identify any materials that have been expended and need to be reordered. What are some easy ways of doing that? A simple way of maintaining a continuous inventory is to have the students identify all the parts of each kit on library cards. Color one side of each card red. Glue a library card pocket on the kit. When anyone has used the kit, the red side of the card is placed outward in the pocket to alert the next person that the kit has been used. Of course, this does not help to keep the kit replenished, but this system does alert the next student to check the kit before using it.

The above idea can be extended by placing two differently colored cards in two card pockets on the kit. In one pocket place an inventory list, the date when it was last checked, and the name of the person who checked it. This list would include all permanent items — e.g., thermometers, containers, and balances. On the second card placed in a separate pocket list the expendable items — e.g., spirit masters, handouts, charts, cotton, alcohol, and seeds. When some item is used, immediately order a replacement, noting on the card when it was ordered. Try to keep the science kits restocked at all times. It takes very little time after an activity is completed to inventory, restock, or reorder; but it means that when you want to teach science, you will not be thwarted because the materials are not in place or have not been replaced.

There are probably other methods that you have used or have seen used by your colleagues. Incorporate these ideas into your materials management systems. The anxiety of not having the materials when you want them can be totally eliminated if there is coordination among the staff members of a school and if the students are given responsibility for specific tasks.

Preparing for a Science Lesson

Another major cause of teacher distress is the extra preparation for science classes. Actually, preparing to teach a lesson is simply one of the "occupational hazards" of the profession. Every *effective* teacher prepares for each lesson in every subject or discipline. For the elementary school teacher this may mean from 10 to 12 different preparations each day. The number may vary among individuals, but the magnitude is there; each planning and preparing step must be carefully accomplished so that your time is used most efficiently.

To conduct a science lesson you will, of course, have to read the teacher's manual and the student guide to understand the learning objectives and the activities to be utilized. You do that with most of your lessons initially, so nothing new is added to your workday.

Now, here is a useful tip. Why don't you and one or two of your colleagues get together for lunch and chat about the science program? Why don't you suggest that they form a "science teaching cooperative" for the purpose of helping each other with the science lessons? Such an endeavor works best where there are two or more teachers assigned to teach the same grade level in the same building. If this is not the case, then coordinate with the teacher of the next grade to form a two-grade-level science co-op. If your classrooms have "open space" designs, this type of co-op will be easy to implement.

The basic idea is to learn and teach only half of the total science program. If your co-op is multigrade level, then you will have to learn a second component. You and your partner(s) will have to make the decision. If you make the decision to team, then each of you takes responsibility to instruct alternating science lessons. With one person acting as the "teacher," while the other acts as the "assistant," you double the available adult help.

Whether you teach science alone or as part of a team, the following suggestions apply. Since most science programs provide a teacher's guide, speed read it as the first step in the overall process of lesson preparation. And because most of you use planbooks of some type, my suggestion is to coordinate the planbook, teacher's guide, student activities, lesson objectives, and evaluation procedures into one total instructional plan. Make notes in the margins of these various items. Keep notes on what works well and what tends to fail. "Debrief" each lesson with just a sentence or two on how you felt the activity went. As you conduct the science program, you will be creating a well-documented file on how to do it better the next year. Keep all your notes and continue to refer to them in the future.

At the same time that you clarify management acts, course goals, and specific learning objectives, you must also identify the student entry levels. What they have already learned, the degree to which they have retained this learning, their motivations, their apparent abilities, and their social and cultural backgrounds are all important data. By knowing several bits of information about your class, you can easily begin to accommodate students with handicaps or learning difficulties in your classroom. When you are aware of special student needs, you can more effectively adapt the physical or instructional environment to parallel unique learner needs.

If you are using a sequential science program, then you will have a general idea of what learnings have preceded your instruction and what will follow. However, it is the students who learn, not the scope and sequence charts, so you must seek information about the kinds of problems that your students might encounter with the present science lesson. Recall those 13 major scientific processes that were discussed in Chapter 1. Many of them are related to learning processes in mathematics, language arts, social studies, and reading. By observing your students during those lessons, you will be able to make some predictions about the ease or difficulty that each one might have with the currently assigned science lesson.

It is critical that you know the relative reading difficulties of the science materials and the general level of reading achievement of the class members. If a small group of your students is below "grade level" in reading, you may have to establish some tutorial groups. Another alternative is to request from the science consultant materials that these children can comprehend. A third technique is to have a good reader prepare audio cassettes so that the poorer readers can listen to and thereby follow the written assignments.

The majority of activity-oriented science programs have rather limited amounts of reading matter. Such programs tend to require each student to synthesize or apply the various cognitive skills that are constantly being built or reinforced in the total school program. It is doubtful that they will have severe reading problems with activity-oriented science programs.

However, you will encounter the problem of students who have difficulty expressing their observations, computing their findings accurately, or preparing tables or figures that display their findings from a selected activity. Again, here is an opportunity to reinforce other skills by using the concrete aspects of the activity-oriented science program. Establish small groups, so that each group must report its general findings to the class, but only after each group member has had the opportunity to present his or her findings to the small group. Think of all the practice in self-expression that the students get in this manner. Rather than the usual question-and-recitation period, the science period becomes one of high personal involvement with a great deal of interaction.

Some of the above tips might not be totally classified as "pure planning." Yet, it is by planning and making contingency plans that the best teachers carry to fruition their instructional goals and objectives.

Other Coping Skills

Once the general plans have been prepared, you learn to cope with the reality that *you* truly do not know all the possible responses to all the possible questions that will be asked. Again, recall from the first chapter that this is one of the "givens" of an activity-oriented science program. Actually, it is a given in science per se. I recently had the fortunate opportunity to work with a group of elementary school teachers through an in-service project which, among other activities, included presentations by university research professors. Each professor gave a general presentation about the kinds of research activities in which she or he was engaged. In 10 separate instances, the elementary school teachers asked these highly published and, in several cases, "world-class" researchers questions to which each and every one politely responded, "I don't know," or, "That would make a great research problem," or, "I'd never thought about that."

Of course, the classic response to your students ought to be, "How can we find out?" You need not burden yourself with the responsibility to know everything there is to know about science. Shift some of that burden to the class members. If any one of them has the interest, you can bet that there will be an oral or written report to the class that addresses the problem. If no one follows up on the problem or the question, then you need not worry either, for that is just what happens in the real world of science: There are questions that go unanswered because they are beyond our current knowledge base or they are too trivial on which to expend one's energy.

One activity that you might consider establishing is a list of "Unanswered Questions" on the bulletin board. Every time someone asks a question that is "unanswered," it is placed on the list by the asker. In this fashion nobody loses face; everyone is a winner. As time wears on, however, there will be students who will discover a possible response to some of these questions.

Another coping skill is to establish a file of recurring problems or questions that tend to be raised each year. Begin that file in your lesson planbook, and add to it after each science lesson, if necessary. At the end of the year or semester, prepare a short recapitulation of the specific problems. Then you can seek the responses from the district's science consultant — if you have one — or from other professionals.

One major coping skill that needs to be sharpened when teaching activity-oriented science is the ability to accept divergent student responses to either questions or interpretations of data. You must expect that in many science lessons, there will be a multiplicity of responses and even some creative ones. One goal of real science is to allow and encourage novel solutions and creative responses.

You must also remember your professional obligation to accept novel student responses. *You* are not placing any *value* on the question or the response. *You* are simply accepting the prerogative of the student to be novel or creative. You may not use subtle "put-down" tactics, regardless of how seemingly outlandish a student's point of view may be or how opposite it is from your expectations. By allowing, nay encouraging, diverse responses, every member of your class can become a star for a moment. That tiny accomplishment is a significant event in the school life of every student.

Several of the above topics may not seem directly related to preparing for science teaching. Yet, keep in mind that you must now be sharpening some special skills that *are emphasized* when teaching science because science is the essence of inquiry. The above ideas follow a logical extension of that essence. In science, one prepares to be just a bit more novel. Or, maybe, it is simply that you will be using many of the skills and techniques that were taught in science courses, but that went without notice. Too, the above methods were probably discussed in teacher preparation classes, but at the time, they may not have seemed relevant to the real world. Whatever the case, teaching science is just that — teaching.

Understanding More About Inquiry

The concept of inquiry is rather difficult to define in nonoperational terms — i.e., without giving precise examples of teacher behaviors and the concomitant student behaviors. As you develop a spectrum of inquiry teaching options, you will better understand the operational meaning by example. Inquiry processes are those that require a high degree of interaction among the learner, teacher, materials, content, and/or environment. Perhaps the most critical aspect of inquiry is that, as it is defined in the dictionary, the student and teacher become consistent askers, seekers, interrogators, questioners, ponderers — ultimately leading to that question that every Nobel Prize winner asks: "I wonder what would happen if . . . ?"

Of course, I don't expect you to win Nobel Prizes, although it would be delightful to see a few of your students do so. What is important is that *you* as the classroom teacher set the stage for the process of inquiry to take place. *You* decide how much time will be spent developing the many processes associated with inquiry behaviors. *You* decide to try another way to teach selected or appropriate units of instruction that lend themselves to inquiry processes. *You* are the one who will systematically teach *your* students how to ask questions. In short, *you* make the difference.

As I mentioned previously and will amplify in Chapter 3, questioning plays a critical role in both the teaching and the learning acts associated with the inquiry mode of learning. Questions lead to investigations that attempt to solve a well-defined aspect of the problem. Such investigations are common to *all* areas of human endeavor. The investigative processes of inquiry involve the students not only in questioning, but also in formulating the question, limiting it, deciding on the best methods by which to proceed, and then conducting the study.

Lest you be a bit dubious about this technique, let me assure you that inquiry techniques provide you with additional options that can expand your own teaching styles — note my use of the plural, *styles!*

Inquiry really is an old technique of teachers. Socrates, Aristotle, and Plato were all masters of the investigative processes. One could argue that the processes that they used have since affected the way most humans in Western civilization think. That heritage has given us a mode of teaching in which students are vitally involved in the learning and creating processes. It is through inquiry that new knowledge is discovered. It is by actually becoming involved in these processes that students become historians, scientists, researchers, etc. — even if for only an hour or two in *your* class.

Using Unguided Inquiry

In the previous chapter, about *guided* inquiry, you noted that the teacher plays the key role in asking the questions, prompting the responses, structuring the materials and situations, and, in general, being the major leader of the learning. Guided inquiry is an excellent method by which to begin the gradual shift from expository or deductive teaching toward teaching that is less structured and more open to alternative solutions. When you sense that the class has mastered the techniques of guided inquiry, then you ought to introduce situations that are still predicated on inductive logic but are more open-ended. In these situations the students take more responsibility for examining the data, objects, or events.

The basic processes of observation, inference, classification, communication, prediction, interpretation, formulation of hypotheses, and experimentation are all a part of *unguided* inquiry. Your role is minimized, which causes a concomitant increase in the students' activity. Let us briefly summarize the major elements of unguided inductive inquiry.

1. The thought processes assume that the learners will progress from specific observations to inferences or generalizations.
2. The objective is to learn (reinforce) processes of examining events, objects, or data, and then arrive at appropriate *sets of generalizations*.
3. You control only the materials and simply pose a question: "What can you generalize from . . . ?" or, "Tell me everything that you can about X after examining these. . . ."
4. The students interact with the stuff of the lesson and ask all of the questions that come to mind without further teacher guidance. Meaningful patterns are generated by the students through individual observations and inferences generated by all members of the class.
5. Materials are essential to making the classroom a laboratory.
6. There is *usually* an unlimited number of generalizations that the learners will generate.
7. You encourage a sharing of the inferences so that all students communicate their generalizations with the class. Thus, others may benefit from unique perceptions.

Unguided inquiry provides a mechanism for greater learner creativity. Further, the learners begin to approach a genuinely authentic "discovery" episode — i.e., the learner finds out something by himself or herself and comes to know that facet.

When you utilize unguided inquiry episodes, a new set of teacher behaviors must appear. You must now begin to act as the *classroom clarifier*. As students begin to make their generalizations, they will predictably make gross errors in logic, state their generalizations too broadly, infer too much from the data, assign single cause-and-effect relationships where there are multiple relationships, and assign cause-and-effect relationships where none exists.

Thus, you must patiently examine the learner in a *nonthreatening* manner to verify the conclusion or generalization. If errors exist in the student's logic or inferences, point these out. But you should not tell the student what the correct inference is, for this would defeat the purpose of any inquiry episode. Your job then is to ask prompting and probing questions.

I suggest that during initial unguided experiences the students work alone. When students work alone, they tend to do most of the work themselves. When they work in pairs or triads, one of the group usually takes the leadership role and dominates the thinking of the group, so that there may be one participant and one or two observers. When students demonstrate the aptitude to use the method successfully in an unguided fashion, then small groups can be assigned to work together.

Some Unguided Inquiry Suggestions. What are some tested ideas that can be used as prototypes as you seek appropriate learning experiences to incorporate into an ongoing lesson? Below are but a few activities.

1. Collect several dandelion flowering heads. Count the number of parachute seeds on each. Make a graph of the results for each flower.
2. Measure the height of some local weed as it grows in the shade and as it grows in the sunlight, and compare those findings.
3. Count the number of leaves on the limb of a small tree.
4. Count the number of vehicles that pass a specific point during a given hour. Then recount the number of vehicles that have only one person — the driver. Then count the number of vehicles by types — large cars, small cars, trucks.
5. Tabulate the highest and lowest temperature readings for several cities for a month or even a year.
6. Count the number of persons who pass a given point in the school lunch line.
7. Tabulate the exact number of minutes that commercials appear on given television shows.
8. Count the number of windows that can be seen in all the houses on one block.
9. Tabulate the number of clear-sky days your town has. Share this information with some school in another city in a different part of the country.

Quite obviously, there are any number of tasks that your students will think of when using unguided inquiry. It matters little if the activities differ from those listed in the science textbook or program that you are using. The important aspect is that the students are identifying and classifying all kinds of objects, facts, and observations. We want them to practice, practice, and practice the processes of inquiry in all phases of their school life. The creativity and spontaneity of your students are the *only* limiting factors. By locating materials from your geographic area, you can build a file of appropriate objects, events, or artifacts that can be used to supplement the usual classroom instruction.

Again, a word of caution. A teacher does not simply jump into inquiry. *You* plan for experiences that are meaningful and relevant and that strengthen inquiry skills in general.

Ultimately your students will begin to make different generalizations and conclusions. One of the generalizations that will quickly emerge as you use unguided inquiry is that there is no *one* way to classify or analyze anything. Your students will arrive at usable schemes that you might not have thought existed.

These exercises can lead to some elements of evaluation. Evaluation is conducted via explicitly stated criteria. But how do we classify or evaluate the criteria? Utility has to be one condition. There are others, but allow the students to determine them.

Problem Solving as Inquiry

A curricular model that could have had even greater impact on the schools was advocated by none other than John Dewey. Among his major educational contributions was the advocacy of a curriculum *based on problems*. He defined a problem as anything that gives rise to doubt and uncertainty. This theory should not be confused with the so-called "needs" or "interest" theories or curriculum. Dewey did actually have a rather careful and definite idea of the types of problems suitable for inclusion in the curriculum. The problems that Dewey promoted had to meet two rigorous criteria:

1. The problems to be studied had to be significant and important to the culture.
2. The problems had to be important and relevant to the student.

It is apparent that recent curriculum projects in science, mathematics, and social studies all tend to be based somewhat on Dewey's problem-solving approach. Most modern curriculums and even a large majority of textbooks suggest problems that must be solved by students utilizing the so-called "scientific methods." Most of the curriculums that you may encounter will stress elements of inquiry, discovery, and other relationships from the structure of the respective disciplines.

When using problem solving with learners, you must consistently play the "great clarifier" role, always helping the learners to define precisely what it is that is being studied or solved. Problem-solving methodologies focus on the students' problems that are being investigated in some systematic manner. The students set up the problem, clarify the issues, propose ways of obtaining the needed information or data to help reconcile the problem, collect the data, formulate conclusions based on the data, and then test or evaluate the conclusions. In most cases the learners will establish written hypotheses for testing.

Another problem-solving approach utilizes films, or loop films. After the film is viewed, ask the class for suggestions about what has transpired. If the class did not perceive any problem, either reshow the film or ask some probing questions concerning the illustrated event. Teachers who have used films in this manner have had almost no problems in getting the students to respond and prepare their own questions.

With any imagination, teachers at any level can identify films from social studies, literature, reading, art, science, or almost any course that have an element of the unknown associated with them and convert that experience into a problem-solving activity. Once the problem has been established, the *students* construct reasonable hypotheses that explain the event. Usually students hypothesize simply by guessing. But guessing is not enough. Once a hypothesis has been presented, it is the students' responsibility to gather data to support or refute that hypothesis — just like real scientists do.

The Learning Environment. You will provide some information to help the students solve the problems, while simultaneously encouraging them to seek data without your help. Further, the classroom must contain the necessary materials by which the learners can test all the hypotheses through planned experiments. Even though there will be a "three-ring circus" appearance in the classroom, it must be stated emphatically that the activity is *meaningful*, *relevant*, and *purposeful*. Self-confidence, initiative, and responsibility are fostered in such an environment.

Evaluating Student Efforts

Stimulating students to use inquiry techniques will be little or no problem. The problem that teachers face is how to "evaluate" the efforts. With the current rage about *accountability*, it would seem that test scores are the only things that count anymore. Yet the really important results of schooling are those that take years to learn and measure — good attitudes, citizenship, ethics, a sense of moral obligation to humankind, care of property, recognition of the rights of others . . . to list but a few. But such philosophizing will not help you convince parents or school trustees that their children are learning something worthwhile in science. To aid, there are a few extensions that might be applied to the traditional report card. However, in science the report card extension would stress the processes, with just a mere passing over of the content.

I am very cognizant that many of the criteria that will be listed require a great deal of subjective judgment on the part of the teacher. But being subjective is not to be confused with being arbitrary. Subjectivity is that element of judging that requires a knowledge of the criteria and a comparison of the individual's performance judged by those criteria. Think for a moment about the judging of a track meet where the criteria are clearly defined for the sprints. Now compare these criteria to those of gymnastics or diving. In the latter areas the criteria are known, but subjectively applied. This analogy applies to the judging of inquiry in the classroom.

Figure 2 presents one of many possible sets of criteria. You must

Criteria for Evaluation in Science	Ranking by Month*			
	Oct.	*Nov.*	*Jan.*	*Feb.*
Curiosity — awareness of environment, questioning attitude				
Initiative — ability to work independently without direct guidance				
Willingness to risk failure to try a novel idea				
Sense of responsibility to the group				
Powers of observation				
Organization and purpose in attacking a problem				
Care and use of equipment				
Recordkeeping — completeness and form				
Communication — relevancy of message				
Ability to classify information				
Ability to formulate generalizations				

* The attitudes and behaviors listed are evaluated on a 1–5 scale, with 1 indicating "Not Usually Observed," and 5, "Always Observed." Criteria that will be evaluated are those having special relevance to the current science objectives and program.

Figure 2. Evaluation of Pupil Progress in Science

select the criteria that you think are *appropriate* to your science classes and then apply them to each student. Over a period of months, you'll observe a definite pattern of science process behavior that indicates the relative growth from using inquiry.

In Conclusion

Quite obviously, there are many other evaluation models that could be illustrated. It is my intent to show simply that evaluation of science processes requires measures somewhat different from the traditional measures of cognitive abilities. Teachers do have anxieties about evaluating students in their classes when they must use "processes." The previous discussion might have given you a few pointers from which to proceed. I would also suggest that you convene an ad hoc committee to study the general problem of science evaluation and that you and your colleagues prepare sets of materials for your school district that would be apropos to the science program being used. After all, the essence of science is involvement. What better way to lose a fear of science than by getting involved in it.

CHAPTER 3
USING QUESTIONS AND TEACHING SCIENCE

The single most common teaching method employed in the schools of America and, for that matter, the world seems to be that of *asking questions*. It may have formally started with Socrates, but the practice still remains first on the list of teaching strategies of all scientists — and science teachers. Therefore, it is important for you to master the logic and application of questioning as you develop your capacity as a highly competent science educator.

Formulating Meaningful Questions

If you are to teach logically, then you must be cognitively aware of the process of framing questions so that student thought processes can be guided in a most skillful and meaningful manner. Implied in the above is that you must design questions that help students attain the specific goals (i.e., performance objectives) of any particular lesson. While textbook and examination questions contribute to the learning process, most questions that occur when teaching science are verbal and teacher formulated.

Although questioning is an important instructional strategy, it appears that teachers may have mistaken the quantity of questions for quality. Researchers have found that some teachers ask as many as 150 questions per class hour. To accomplish such a feat, most of the questions would have to be fact oriented (or very low level). While teachers often stress that thinking is important, their questions do not reflect this emphasis.

There are several reasons teachers support the use of so many fact (low-level) questions. One rationale proposed by many teachers is that students need facts for high-order thinking. This is a cogent point, but there are additional ways to teach facts, such as programmed instruction. Teachers also apparently overuse fact questions because they lack systematic training in the use of questioning strategies. Studies conducted by myself and my colleagues revealed that when teachers are trained in questioning, the frequency with which higher-level questions are used in the classroom goes up significantly.

Another reason that teachers have tended to ask so many low-level questions is that, until recently, they have lacked an easy-to-use system to organize and classify questions. They have also lacked a means of evaluating the effect that different questioning techniques have on the learning process. This is where Bloom's *Taxonomy* can be of use. A majority of textbook authors use Bloom's *Taxonomy* to evalute the potential for critical thinking in the classroom. This question classification system is based on the type of apparent cognitive process required to answer the question. Benjamin S. Bloom and others[2] use six cognitive categories to classify questions: knowledge, comprehension, application, analysis, synthesis, and evaluation. The thinking processes involved in this model progress from the simplest (knowledge) to the most complex (evaluation).

Each higher cognitive process probably includes all lower cognitive processes. For example, if a teacher asks an evaluation question, the student will normally use synthesis, analysis, application, comprehension, and knowledge to answer the question. Again we caution: When building skills and concepts, you may have to devote large blocks of time to knowledge or level-one objectives and questions. But we urge that you then begin to build upward. If teachers emphasize low-level questions when it is high-level questions that stimulate thinking and evaluation, then we may have discovered one of the basic reasons why students find science boring.

Although the evidence is somewhat inconclusive, there does appear to be a direct relationship between the level of questions asked by the teacher and the level of student responses. Further, it appears that if a teacher decides to raise her or his expectations for the class and *systematically* raises her or his level of questioning, then all students raise the level of their responses accordingly. Of course, this implies a carefully planned questioning sequence that would probably span several weeks of instruction.

Tips for the Teacher

The implications of the above for teacher decision making are many. First, if you want your students to develop higher levels of thinking, to evaluate information, to achieve more, and to be more interested, you must learn to ask higher-level questions. Second, you must encourage your students to ask more questions, and more *thought-provoking* questions, if you desire greater student involvement in the process of learning.

An example of *one* form that you can use to evaluate questions is shown in Figure 3. The form as presented is rather detailed, but you can abbreviate it so that it fits on a $3'' \times 5''$ card. The appropriate categories could be summarized on the card and then tallied as they are generated — in the classroom! Or you could tape-record a class session, and then tabulate the questions during a free period. After establishing a baseline rate, which would represent the percentages for the various categories during a specified science period, you could then systematically devise a change strategy to improve any specific cognitive-skill–questioning area.

Another important implication for those of you who desire to stimulate critical thinking is that you should be aware of the advantages and disadvantages of your science textbooks and other printed materials. To obtain the objective desired, you will usually have to supplement the materials provided. One trick to use to develop student comprehension is to ask them to make up two questions about some passage in the science text or about the data that they collected in some related activity.

Summary. In teaching science you will realize that high-level questions demand more science activities. In addition, you will find yourself reducing the number of "right answer" questions, with a concomitant increase in the number of open-response questions. And this is just what you want to do!

Applying Three Strategies

The preceding discussion should have given you a better appreciation of the role that questioning plays in stimulating student

Category	Expected Cognitive Activity	Key Concepts or Terms	Sample Phrases and Questions	Tally Column	% of Total Questions Asked
1. Remembering (Knowledge)	Student recalls or recognizes information, ideas, and principles in the approximate form in which they were learned.	memory, knowledge, repetition, description	1. What do the charts show…? 2. Define…. 3. List the three…. 4. Who invented…?		
2. Understanding (Comprehension)	Student translates, comprehends, explains data and principles in her or his own words.	explanation, comparison, illustration	1. Explain the…. 2. What can you conclude…? 3. State in your own words….		
3. Solving (Application)	Student selects, transfers, and uses data and principles to complete a problem task with a minimum of directions.	solution, application, convergence	1. If you know A & B, how could you determine C? 2. What would happen if…?		
4. Analyzing (Analysis)	Student distinguishes, classifies, and relates the assumptions, hypotheses, evidence, conclusions, and structures of a statement of a question with an awareness of the thought processes being used.	logic, induction & deduction, formal reasoning	1. What was the purpose of…? 2. Does that follow? 3. Which are observations, and which are inferences?		
5. Creating (Synthesis)	Student originates, integrates, and combines ideas into a product/plan that is new.	divergence, productive thinking	1. Make up…. 2. What would you do if…?		
6. Judging (Evaluation)	Student appraises, assesses, or criticizes on the basis of specific standards and criteria.	judgment, selection	1. For what reason would you favor…?		
			TOTALS		100%

Figure 3.   Classroom Question Classification Method for Science Teaching

achievement in science, as well as of the crucial effect that your questions have in encouraging higher-level cognitive processes in your classroom. As an operating procedure it can be generalized that the type of questions you ask will be viewed by the students as indicating the types of learnings that are important.

For your convenience, all science questions — whether asked by teacher or student — might be classified into three categories or patterns: (1) convergent, (2) divergent, and (3) evaluative.

Convergent Questions. As the term denotes, the convergent questioning pattern focuses on a rather narrow learning objective, utilizing questions that elicit student responses that *converge* or *focus* on some central theme. Convergent questions for the most part elicit rather short responses from students — e.g., "Yes" or "No" or very short statements — which tend to be at the knowledge or comprehension level. The use of the convergent technique per se is not to be construed as being "bad." There are many situations in which you might desire the students to demonstrate a knowledge of specifics, and thus lower-level questioning strategies would be most appropriate.

For example, *if* you use an inductive teaching style (proceeding from a set of specific data and aiming for student conclusions), *then* you utilize a large proportion of convergent questions. Too, you may wish to use short-response questions as warm-up exercises by which to open, close, or break the monotony of the traditional classroom, perhaps using a "rapid-fire" approach. This technique would be most appropriate where you are building vocabulary skills; keep in mind that much of science is initially learning a kind of foreign language.

The use of a rapid-fire convergent technique also allows for participation by a very large number of students as you focus on specific learning objectives, skills, terminologies, or solutions to easily solved activities having a specific "answer."

Below is a list of convergent questions. Note that these questions all meet the criterion of focusing the student responses on a narrow spectrum of possible options, and that responding students will rely on recall more than analysis.

1. How could heating a tin can with a little water in it and then capping it possibly cause the can to collapse?
2. Under what condition does water expand?
3. What effect does acid have on blue litmus paper?
4. Where do you find the tallest dandelion plants?
5. What are the Van Allen Belts?

Patterns for Divergent Thinking. Divergently oriented questions seek responses that lead to a *set* of foci or a *spectrum* of responses. Divergent questions also elicit longer student responses.

When the results of experiments or science activities are being discussed and you want to elicit an array of student responses, divergent types of questions are appropriate because multiple responses almost always occur. You can capitalize on this by asking a question that can be answered with multiple responses, calling on three or four students in turn, and then assuming a passive role in the discussion; this, as you know, is a rather sophisticated strategy. And because divergent questions generate a multiplicity of responses, you must be prepared to accept all student responses, to allow or encourage novel solutions and creative responses.

One technique that will aid you initially in framing divergent questions is to write the questions on paper prior to asking them. In this manner you can examine them to ensure that they are clearly stated and convey the precise meaning you intend. The first time you use divergent questions, you may find the class experience rather difficult or even disappointing. Because many teachers still devote the majority of time to the oral recitation of very low-level learnings, students may not be oriented toward providing responses that are longer and/or that result from higher-level thought processes. It takes a good deal of reshaping of student behavior patterns to elicit proper student responses through divergent questioning techniques. But alert the students that the questions will be varied and have patience. By using divergent questioning you will soon discover that your students are dealing in the higher-level thinking categories of the cognitive taxonomy — i.e., application, analysis, and synthesis.

Science by its very epistemology — the experiment or activity — automatically creates diverse sources of information. Use those sources to create wider viewpoints in the class. Below are selected questions, adapted from the list previously presented as convergent questions, that may now be classified as divergent.

1. What types of structures are collapsed when they are heated with a little water in them and then have the opening capped?
2. What would happen if water contracted rather than expanded on freezing?
3. How many ways could you test to determine if you have an acid?
4. How does the environment affect the early development of young seedlings?
5. Why would one want to know about the Van Allen Belts?
6. If we exhaust the nation's petroleum resources within 10 or 20

years, what will be the impact on our standard of living?
7. List as many alternative fuel sources as you can, other than gasoline.

The Evaluative Questioning Pattern. The third basic pattern of questioning is one that utilizes divergent questions, but with one added component — evaluation. The basic difference between a divergent question and an evaluative question is that the latter has built within it an evaluative or judgmental set of criteria. When you ask *why* something is "good" or "bad," you are raising an evaluation question. However, because it is possible that an evaluative question might elicit nothing more than a poor collection of uninformed student opinions, you must emphasize the criteria by which a student renders a judgment. These criteria should concern the worthwhileness or the inappropriateness of an object or an idea.

You can systematically help students in developing a logical framework by which to establish evaluative criteria. For example, if you ask a question and the student replies with "Because," then you will recognize that the student is lacking in logical perception, may be dogmatic or arbitrary, or just does not understand how to go about framing a logical and consistent set of evaluative criteria. Once again, we caution you not to use sarcasm or any other put-down technique; the typical teacher comment that "You're not being logical" gives the student no basis for improvement whatsoever. Take a positive approach and reinforce the student by providing examples that yield a logical development of evaluative criteria. Provide a specific set of criteria or specific items so that the student develops his or her own specific set of criteria. In this manner, a student will understand why value judgments and opinions are being held.

Observations tend to verify that as evaluative questions are presented and student responses elicited, the teacher and the students tend to want to classify the evaluative responses along some type of continuum ranging from "bad" or illogical responses to "good" or logically developed responses. How do you classify evaluative responses? By logical development, internal consistency, validity, and perhaps responsibility. In short, it is suggested that you once again accept all student responses and that you discuss apparent logical inconsistencies that develop after the student has had an opportunity for classroom discourse.

Following are examples of evaluative questions. Remember, most, if not all, evaluative questions will be divergent; the one criterion that separates divergent questions from evaluative ones is that the latter rely on the establishment of judgmental criteria or the judging of the value of some idea based on other pre-established values, criteria, or conventions. Note that some examples previously designated as divergent questions have been converted below into evaluative questions.

1. Why is Newton's Third Law of Motion — "For every action there is an equal and opposite reaction" — so important in our lives?
2. Why is the world a better place in which to live because of computers?
3. React to the following newspaper headline: "Inventor Claims Perpetual Motion Machine."
4. How has the federal system of interstate highways harmed or helped our environment?
5. Defend the strip mining of coal in eastern Montana.
6. To what extent is the gene theory more viable than that of spontaneous generation?

Did you realize that in discussing divergent and evaluative questions, I have been using the term *responses*, not *answers*? Answers carry the connotation of being complete or the single absolute final word. To be sure, convergent questioning patterns may elicit student *answers*. However, you must recognize that when divergent and evaluative questions are framed, the students will be providing *responses* — the degree of finality is not there.

Technical and Humane Considerations

Framing the Questions. The use of classroom questions in a science period, a tutorial period, or an inquiry session is always predicated on the assumption that meaningful or purposeful learning activity is taking place. To ensure that this occurs, you must ask questions in a positive reinforcing manner. That is, questions should be used so that the student enjoys learning about science and will receive positive reaction to his or her responses.

The basic rule for framing a question is: *Ask the question; pause; call on a student.* This rule is grounded in the psychological principle that when a question is asked, and then followed by a short pause, all students will "attend" to the communication. The nonverbal message is that you might ask any student in the class to respond. Thus, the attention level of the class remains high. If you reverse this pattern by requesting a particular student to respond prior to actually asking the question, then all those students who are not involved have an opportunity "not to attend" to the communication between you and the student. This same type of question framing can be used even when employing the multiple response technique wherein you request several students to respond.

When you develop the habit of pausing after asking a question, you will learn not to "dread" the wait time. Mary Budd Rowe discovered that teachers are most impatient with students when asking science-related questions.[3] She measured the "wait time" of many metropolitan school teachers and found that wait time between asking a question and either answering the question before the student or calling on another student had to be measured in *fractions of seconds!* Is it any wonder that some students dreaded to be called upon. They knew it would elicit impatient behaviors from the teacher. I want you to know that classroom silence is not bad when asking questions and waiting for the responses.

Probing Techniques. Once activities have been concluded, a question has been asked, and a student has been identified to respond, there is always the possibility that the student will not answer the question *completely*. This is a common occurrence in science classes. When this does happen, you need to move into probing or prompting strategies that attempt to clarify the question so the student can understand it better, that cause the student to amplify the response, or that elicit additional responses from the selected individual so you can verify whether or not the student comprehends the material.

If a student has neither clarified the question nor adequately amplified the response, then use a probing technique. To do this in a positive manner, acknowledge the attempted response but then encourage the student to clarify or amplify it.

Prompting. During a science discussion you must prompt students so that an incomplete response can be transformed into a more complete or logical response. Basically you will use the same technique as discussed above regarding probing — i.e., you will always aid the student with a positive reinforcement so that the student is encouraged to complete any incomplete response or revise an incorrect one. In most cases the student will respond to a question with a partially correct response. Or stating it negatively, a student will often respond with a partially incorrect response in addition to a partially correct one. Immediately upon hearing the response that fits this category, begin to prompt the student so that the response can be completed, made more logical, re-examined, or stated more adequately or more appropriately.

Handling Incorrect Responses. No matter how skillful you are in motivating students, providing adequate and relevant instructional materials, and asking meaningful questions, there will be one continual problem that detracts from the intellectual and interpersonal activities of a science lesson — incorrect student responses.

As was discussed previously, you can use probing and prompting techniques when a student response is partially correct or incompletely stated. Basically, prompting and probing are rather easy techniques because you can reinforce the positive aspect of the student's response, while ignoring the negative or incomplete component. However, when a student gives a totally incorrect response, a more complex situation arises. First, you have little to reinforce positively, and such teacher comments as "No" or "That is incorrect" act as negative reinforcers which, depending on the personality of the student who responds, might reduce her or his desire to participate in science discussions or recitation. Second, if you respond very adversely to an incorrect student response, there is a high probability that the "ripple effect" will appear. Jacob S. Kounin[4], who has so described this effect, demonstrates that students who are not themselves the target of the teacher's aversive strategy are, in fact, affected by what the teacher does to other class members.

What, then, can you do? Since the entire approach to this method is to stress the positive, the first decision you might make is whether any portion of the student's verbal response can be classified as valid, appropriate, or correct. Following this "split-second" decision making, you must then provide positive reinforcement or praise for that particular portion. When an incorrect student response provides no opportunity for positive reinforcement, you might attempt to move to a *neutral* probing or prompting technique. For example, you might state, "Your response is in the magnitude of the answer," or, "Could you tell us how you arrived at your answer?" Note that none of these responses is totally negative, but each can be considered as neutral in that it is not positive either.

Concept Review Questions. As you begin to develop confidence in using various questioning techniques, it becomes necessary for you to review in a most efficient manner those previously learned science concepts and to relate or correlate them to knowledge that will be generated at a later date. In most cases, teachers tend to schedule a review prior to a summative evaluation. "Review Thursday" tends to be an ineffective use of student time in that the vast majority do not need the review, and for those students who do, such an oral exercise is usually fruitless in expanding their intellectual understanding of whatever it is that the teacher desires.

How can you review previous concepts while conducting questioning strategies? One successful method is to re-enter concepts discussed previously but in the context of newly presented material. For example, if you are progressing through a unit on electrical currents, and the topic of batteries has already been covered, and a related topic — e.g., chemical reactions that produce energy — is being studied, you begin to ask questions relating to batteries and their chemical reactions. It's that simple.

The concept review technique requires that you be always on the alert for a teachable moment that will allow a meaningful relationship to be established, a previous concept to be reinforced, or a synthesis of knowledge to take place, thereby creating one more motivational factor for the class.

Encouraging Nonvolunteers. With most science activities, you will have no problem encouraging students to respond to questions or to present their findings to the class. But what are some helpful strategies in motivating nonvolunteers to respond verbally during a questioning session?

The first technique is to *maintain a highly positive approach toward the student*. Once the nonvolunteer has responded appropriately, there should be generous positive feedback to encourage the student to continue such behavior. Another technique is to ask rather easy evaluative questions since most students respond to questions that concern judgments, standards, or opinions.

Another method that can be used to increase nonvolunteer participation is to make a game out of questioning from time to time. Place each student's name on a card so that you can draw the cards at random, thus creating a condition where every student could be called on to recite.

Too, there is nothing wrong with giving the known nonresponding students a card with a question on it the day before the intended oral recitation period. Very quietly hand these students a card and tell them they might check over the assignment so that they can summarize their responses for the next class period. At least this method begins to build a trusting relationship between you and the students.

I do not condone calling on nonvolunteers for aversive or punishment tactics. Schooling ought to be positive with affective consequences of "approach tendencies" being emulated. As a general rule the most influential means by which you can encourage a nonvolunteer to participate is to be sincere and genuine in treating each student as a human being. Nonvolunteers have learned, and probably painfully, that it doesn't pay to say anything in the class because the teacher will "put you down" anyhow.

Developing Student Skills in Framing Questions

The previous techniques have been oriented toward the teacher, but there is another source of questions that is often overlooked — the students themselves. Classes can be organized to encourage student communication, giving each one a chance to express opinions and ideas; but evidence shows that teachers do most of the talking.

Studies have shown that (1) students can be encouraged to ask productive or higher-level questions, (2) the more questions a student asks per period, the greater the probability that the questions will be at higher levels, (3) praise will encourage and stimulate more productive thinking processes among students from lower socioeconomic backgrounds, and (4) students become more involved in the classes in which they are encouraged to ask questions.

To teach students how to frame their own good questions, refer to the game that was first made famous many years ago on radio, "Twenty Questions." The game of "Twenty Questions," in which participants ask questions to identify something, can be applied in the classroom. You can present a problem or identify some concept that needs to be discovered and then allow the discovery to take place only through student questioning. Initially you conduct the session. But as students become more proficient at questioning skills, then they might conduct the entire session with you merely analyzing the various interpersonal reactions.

J. Richard Suchman[5] prepared an *Inquiry Development Program* published by Science Research Associates where the emphasis is on developing student questioning skills. Using this approach, you present a problem to the students and then play a passive role in the learning, responding only with a "Yes" or a "No" to any student's question. What this means is that the students must learn how to ask questions on which they can build a pyramid of knowledge, ultimately leading to a convergent response or answer, rather than simply a series of unsystematically asked questions. When this technique is utilized with students who have had almost no opportunity in the classroom to ask questions, the initial results can be rather "sad." However, you should review each lesson and give precise and detailed directions on how the questioning can be improved. As one alternative, if it would not be too slow, you might initially write each student's question on a chalkboard or on an overhead projector transparency so that each student would have visually presented those questions that are asked by her or his peers. In this way information and skills can be built up gradually in a somewhat systematic manner.

Of course, when developing student skills in framing questions, it becomes imperative that the students understand the logic that each question must encompass a large category of specifics. In short, you must give practical application to deductive logic skills.

Another alternative to use in developing students' skills in framing questions is to have students prepare recitation questions based on the science data being studied. In this manner you can assign a few students each day to prepare a series of questions for their peers. To be sure, most students will be oriented only toward facts since that is what is most often reinforced in their learning. But if you are skillful in continually reinforcing those questions that are aimed at higher-level thinking skills and in ultimately helping each student to prepare appropriate higher-level thinking questions, you will gradually observe improved questioning skills.

As a teacher you will note that as you begin to encourage the class members to ask questions of each other, there is a subtle shift of responsibility to the class. Teachers usually admonish their charges to "accept more responsibility." I submit that it is in the learning situations where greater responsibility may be acquired. The latter statement implies that responsibility is a "learned behavior" just like so many other behaviors. As a teacher you owe it to your students to help them become more articulate and thoughtful individuals. What a splendid opportunity exists by a slight shift in classroom questioning techniques. As was stated previously, this technique must be carefully explained to the students and then practiced for a few class periods. You and the class can generate a set of criteria that provides information or rules by which the various student-made questions are framed. Then, perhaps once a week or more often, the students can conduct the questioning sessions. Further, this method acts as a prerequisite experience to student-led discussions.

Teachers with whom I have worked have all been pleased with the results of such techniques. And, more importantly, these same teachers were amazed at how they *underestimated* the potential that existed in their classes. I am not implying that these techniques are simple to implement. It takes much work and planning. But the attendant rewards make both teaching and learning more worthwhile.

Teacher Idiosyncrasies: A Caution

One would speculate that all teacher behaviors that may be associated with questioning are positive and encouraging. After all, one needs only a few tricks and a smile and *shazam* — instant success! Unfortunately, there are teacher behaviors that, when used inappropriately, may interfere with a smooth classroom verbal interaction pattern. Briefly these idiosyncrasies are (1) repeating the question; (2) repeating all the student responses; (3) answering the question — yes, answering one's own question; (4) interfering as a student completes a long response by cutting the student off; (5) ignoring — i.e., not attending to — the responding student; and (6) calling on the same few students.

Summary

If you have perceived that I am attempting to make a "game" of science teaching, then you are absolutely correct! Science ought to be a time in which everyone has fun learning or at least will have *approach tendencies* while learning. If science can be meaningful and relevant, then students will enjoy *working* at it. Students have approach tendencies for those areas in which they are successful; if science is distasteful, then it is because, for the most part, students have been unsuccessful in science. Such an attitude can be easily rectified by making science, or for that matter all subjects, success oriented.

All of the above questioning strategies must be considered among your tools of the trade. But learning would soon become a very trite and boring exercise if it were centered only about questioning. While questions comprise an important set of teaching tools, they are just that — tools. Each technique must be used appropriately and must be congruent with the objective that you have for any specified student, group of students, or class.

Finally, you might realize at this point that the techniques you need to apply to improve the teaching of science are some of the same that can be used to improve all subjects.

CHAPTER 4
BEING SUCCESSFUL WITH SCIENCE-RELATED DISCUSSIONS

The most successful elementary school teachers tend to use a mix of individual and small-group work. Both of these topics have been previously introduced. Thus, those promised elaborations with the emphasis on *process objectives* rather than *performance or behavioral objectives* are begun. The latter type of objectives specifies exact learner behaviors in terms that make learning observable. Nearly every teacher in America knows how to identify, write, and specify learning in performance terms.

However, there is yet another type of objective that may be of equal, if not greater, importance — *the process objective*. A process objective requires the learner to participate in some technique, interaction, or strategy. Process building is much more subtle than specifying performance objectives and requires that you carefully plan *experiences* for the learners. As was mentioned previously, the schools should develop responsibility, a process which takes years to accomplish and which some individuals never master. Development of a process such as responsibility requires that the students have something about which to be responsible. Likewise, practice, planning, and cumulative experiences are all necessary to develop successful processes in small-group and discussion techniques.

It may be confusing to use *process* with two definitions: (1) the 13 scientific processes, and (2) the process objectives that relate to learner interaction with techniques or strategies. So, to avoid any problem with semantics, I will always refer to the first class of processes as *scientific processes* and the second classification as *interaction processes*.

On analysis you will discover that the scientific processes tend to belong to the cognitive domain of skills, while interactive processes tend to fit the realm of the affective domain — the attitudinal dimensions of life and learning. Let us examine how you can determine the status of your classroom group and what organizational and individual developments are necessary to use the interaction processes most effectively.

The Basic Organization

Establishing Goals. The first requirement of conducting a successful small-group science learning activity through either a discussion or a small-group learning unit is the development of a set of *long-range* priorities. While performance objectives are written for the immediate, interaction process objectives are written for the development of skills, attitudes, and knowledge that require long periods of time to develop. Long-range objectives are important because, as the planner, you must identify the skills that the learners must demonstrate before they can approach or achieve mastery of any stated objectives.

As an aid to your long-range planning there are several skills that both you and the learners must be able to demonstrate. These might initially be called *pre-discussion skills*. Below is a listing of both the order and the types of skills that are prerequisite to conducting successful small-group discussions in science:

1. The teacher knows how to select and ask questions in a systematic manner.
2. Students learn to respond to divergent questions.
3. Students learn to respond to evaluative questions.

4. The class is subdivided into small groups to discuss topics that require divergent and convergent responses.
5. Students can complete committee tasks.
6. Students learn to ask questions of each other and of the teacher.
7. The teacher identifies the needed *interaction processes* for specific individuals and for the class as a group.
8. Small-group units are formed in the class.
9. The teacher prepares student leaders, recorders, and observers.
10. The class is subdivided for teacher-led small-group discussion.
11. All class members comprehend the concept of formative evaluation.
12. The teacher and the students plan for appropriate discussions.
13. Students learn roles for various discussion techniques.
14. The class is subdivided for student-led discussions.
15. The entire class critiques and evaluates small-group efforts.
16. The processes associated with small-group discussions are incorporated into the ongoing activities of the class.

On first reading, you might be overwhelmed. Please don't be. The above 16 major tasks are easily incorporated into science instruction, but they are also easily implemented during social studies or language arts lessons. The list is provided to help you evaluate your own classroom and what needs to be done to improve student skills through meaningful experiences.

You already have studied, and perhaps mastered, the art of questioning which was detailed in Chapter 3. More than likely you are already using some elements of individualized learning such as learning centers in your classroom. Most teachers utilize some small-group activities, often through committees. All that I am proposing is that you extend these tested teaching strategies to the field of science in a systematic manner. Let us develop the entire technique.

Developing Small Groups. Group size is an important variable because it influences learner participation levels. To state absolute minimum or maximum numbers to ensure successful interaction is difficult. An optimal size generally appears to range between four and eight.

When four or fewer students are involved in a science discussion group, there is a tendency to "pair-off" rather than interact. Conversely, the likelihood that *all* students will participate decreases when the group number approaches 12. With larger numbers — e.g., 15 or more — a few students tend to remain very interactive, a few somewhat interactive, and the great majority silent or passive. It seems appropriate to subdivide groups of 12 or more into groups of six to eight prior to the initiation of a small-group discussion. The topic, the group, and the leader's experience all affect this decision.

A discussion denotes an exchange of ideas with active learning and participation by all concerned. On the other hand, recitation tends to be a passive technique from the viewpoint of student activity. Discussion methods seem to be logical extensions of student interactions with materials or objects during science activities as student and student or student and teacher exchange thoughts. Discussions allow students to discover, develop, state, and react to personal viewpoints — not merely to repeat those ideas that you or a text has presented.

For purposes of clarification, a science discussion is described as including these elements:

1. A small number (6–12) of students meeting together.
2. Recognition of a common science topic or problem.
3. Initiation, exchange, and evaluation of information and ideas that relate to science.
4. Direction toward some goal or objective (often of the participants' choosing).
5. Verbal interaction — objectively and emotionally.

Why Use Discussions?

Discussions and small-group learning units are most appropriate *if* you desire to increase teacher–student and student–student verbal interaction in the classroom. Recall in Chapter 1 that 12 different goals for using small-group or discussion techniques were described as being appropriate to aid students in becoming more responsible and independent learners. During the actual science lesson, students become active participants in the class activities and demonstrate both their scientific and their interaction process skills.

A well-accepted psychological principle is that students learn best when they are actively involved or participating. If you desire to promote a wide range of interests, opinions, and perspectives in the science class, then small-group discussions provide another way to accomplish that goal. If you desire to have different students doing different tasks or activities at the same time, all leading to meaningful goals, or if you desire to practice indirect control of the class, then discussion is an appropriate technique.

Small-group discussions provide an excellent method for dealing with controversial issues or open-ended questions. The need for a local, state, or national science policy, the problems of advocating a particular solution or point of view, and efforts to inform are all appropriate subjects for science-related discussions.

One criterion is critical in planning for the use of small-group discussion activities: Is the activity, question, or group task one for which there is abundant data? This criterion requires that you know approximately the range, amount, usefulness, and timeliness of available materials. If the students are limited in their research to collections of outdated, inaccurate materials, then they will not be able to challenge prejudices, develop open-minded and flexible approaches to new information, or learn how to reflect upon the acquisition of new knowledge.

It is your responsibility as a teacher to develop the habit of thinking in terms of the-students-and-I-working-together. A "we" attitude helps you in establishing clear goals that revolve around teacher–student relationships, student–student relationships, the learning purposes of the classroom, and a supportive emotional climate and learning atmosphere, so that each class member respects all other individuals and their respective ideas.

Every one of us has experienced, in either large or small classes, how the initial sessions are often marked by a lack of responsiveness and a general climate of anxiety. Group development and cohesiveness are attained only gradually. Effective small-group facilitators understand how to plan experiences that reinforce group goal setting, group effectiveness, group interaction, and group development.

Studies have shown that, when conducted under appropriate conditions, small-group methods are superior for selected purposes. There is evidence that changes in social adjustment and personality can be facilitated through small-group instructional methods. Further, it has been demonstrated that small-group activities help to increase the students' depth of understanding and grasp of course content. Two affective consequences have been demonstrated as being attributable to small-group techniques: (1) the enhancement of motivation and greater involvement by the students, and (2) the development of positive student attitudes toward course materials.

Finally, you will find that when you use small-group discussion techniques, your students will develop science problem-solving skills because they obtain greater practice in the application of concepts and because they realize that the content information has a practical use.

Two general skills — inquiry and cooperativeness — will tend to increase the effectiveness of the small-group participants. As groups develop cooperative members, the quality and quantity of learning become amazingly high. Conversely, if the group members compete with each other, there is a tendency for both the quantity and the quality of learning to decrease. Of course, to reach selected instructional goals, intergroup competition may actually be desirable (if not carried to an extreme). The overall success of small groups within your classroom depends on your selection of a blend of discussion modes, some of which require intragroup cooperation and a few that call for intergroup competition, which may be in the form of science-related games or simulations.

Introducing the Concept of Evaluation

Evaluation of discussions is designed to provide feedback to each individual who participates in the group activity. Since small-group discussions are interaction-process–oriented, that process should be continually evaluated so that each participant may improve. In the preparation of such evaluations, simplicity is the key concept. You ask what the goals or objectives of the activity are, identify appropriate criteria by which to judge each component, and then prepare a model form for the evaluator.

Once the evaluator has judged the group activities, data from each individual should be compiled so that the group can receive aggregate or cumulative feedback. It is possible to tally all of the individual responses for each item and present the sums to the group. This technique allows each individual to compare the self-rating results to that of the group.

Evaluation forms, often simply homemade, may be filed by you or by each student in order to determine the type and direction of growth for each participating individual. Such data enable you to help students who have not mastered specific discussion skills and to provide future direction for the group's use of discussion.

One form of evaluation is simply to record the number of times each student interacts verbally. This form could be prepared without any special printing. It would look like Figure 4. The evaluator tallies a mark each time any individual speaks. At the end of the discussion the leader, and perhaps even you, could examine the tabulation to determine if someone dominated

Names of Group Members	Number of Interactions
1. Laura Anne, Leader	⊪⊪ ⊪⊪ ‖
2. Gus, Recorder	‖
3. Jerry	∣
4. Elaine	⊪⊪ ‖
5. Sharon	⊪⊪
6. Pat	⊪⊪ ∣

Figure 4. Tally Method Evaluation Form

the discussion or if someone did not contribute to it. The rationale for this type of evaluation is to promote interaction behaviors, not to blame anyone.

Another possible model instrument that is designed to get feedback from the participants themselves, not just the evaluator, is shown in Figure 5. You or a small-group evaluator might then compile group data on a graph to better observe the total range of responses.

Other evaluation forms might help group members assess their own participation over time or test the affective dimensions of the discussion (particularly when decisions or value judgments are made or data interpreted). The form in Figure 6 could be used during a science discussion that focuses on value-laden ideas or decision making.

Techniques for the Classroom

Let us now address four well-tested small-group techniques that are most suited to science classes: (1) brainstorming, (2) Phillips 66, (3) tutorial, and (4) task groups. You might even mix some of these as you gain experience with them.

Brainstorming. A very simple technique that is useful when creativity is desired is *brainstorming*. Most science activities have some elements that require students to do some freewheeling thinking. This is when you want to use a brainstorming group. Any number of students can become involved in a brainstorming activity. The shorter the discussion period is, the smaller the groups should be, so let time dictate the size within a 5- to 12-person limit.

The brainstorming session is started by the leader who briefly states the problem under consideration. The problem might be as simple as "How can we collect data about the problem?" or as complex as "How can we set up an experiment to test seed germination?"

After the topic is stated and before interaction starts, it is crucial to select a method of recording the discussion. It could be taped, or one or more students who write quickly could serve as recorders.

Although all the students will be oriented to the rules, make sure that the student leader enforces these procedures. The following rules seem to be especially important:

1. All ideas, except for obvious jokes, should be acknowledged.
2. No criticism is to be made of any suggestion.
3. Members should be encouraged to build on each other's ideas. In the final analysis, no idea belongs to an individual, so encourage "piggybacking."

Directions: To evaluate your group, place an "X" on the line above the statement that best describes your reaction to each of the incomplete sentences.

Group _____ Date _____

Name _____

1. I thought that the discussion

| Gave everyone a chance to participate freely. | Allowed almost everyone a chance to participate freely. | Was dominated by only a few. |

2. As far as my participation in the discussion, I

| Was really with it. | Could have done better. | Was totally out of it. |

3. The discussion leader

| Encouraged a wide range of participation. | Selected only a few persons to participate. | Seemed to dominate the discussion most of the time. |

Figure 5. Checklist for Discussion

Directions: The student who evaluates the discussion should circle the response that describes the conclusion. If any negative evaluations are given, then a short statement of how that aspect can be improved must be given.

1. To what extent was the task clearly defined?	Well defined	Somewhat defined	Needed clarification
2. Were all conclusions definitely stated or identified?	Very well stated	Somewhat stated	Suggestions to improve
3. How would you rate the value of the conclusions?	Very practical	Somewhat practical	Impractical
4. Were the conclusions made in light of the problem?	Yes	No, the conclusions were made by considering other data	
5. How well did the group share information?	Much sharing	Some sharing	There seemed to be a need for more
6. To what extent could you determine if the participants were pleased with the manner in which the discussion took place?	All seemed to be pleased	Most seemed pleased	There seemed to be a mix

Figure 6. Discussion Product Appraisal Form

4. Solicit ideas, or opinions, from silent members. Then give them positive reinforcement.
5. Quality is less important than quantity, but this does not relieve the group members from trying to think creatively or intelligently.

Brainstorming is an initiating process and must be followed up with some other activity. *After* the discussion or brainstorming session, it is important that the ideas presented be classified by types and then evaluated for use by students in follow-up activities. One way to follow up would be to use the ideas generated in the brainstorming session as the basis for another type of discussion. Brainstorming can also lead to the prioritizing of the elements — e.g., when you desire a series of suggested science topics to be assessed in priority for future study.

The evaluation of a brainstorming session should not be lengthy, and it should be nonthreatening for the participants. Remember, you want everyone to contribute, regardless of their level of academic capability. You may want to make some private assessments about academic levels, levels of inhibition, or who is "turned off" by science; but all public evaluations must be highly positive in nature.

Phillips 66. The "Phillips 66" discussion group involves exactly six students, and was developed by J. Donald Phillips at Michigan State University. It is established quickly and does not call for pre-orientation. Students do not have to be highly skilled in group interaction for this type of discussion to work effectively. In fact, the Phillips 66 technique is most appropriate as an initial mixer activity.

The class is divided into groups of six (this can be done by you or on a volunteer basis). The groups then have one minute in which to select a secretary and a leader. At the end of one minute, you give a clear and concise statement of the problem or issue for discussion, worded so as to encourage specific single statement answers. The students then have exactly six minutes to come to an agreement as to the best solution for the problem.

When using the Phillips 66 group in the primary grades, you may decide to eliminate the role of secretary, but I would encourage you to still consider the benefits of having one of the students summarize the group's solution. Listening and summarizing are important skills for group work, and the Phillips 66 method is a good training technique for future group leaders, recorders, and evaluators.

The Phillips 66 discussion group can be very useful as an initiating activity for a concept formation or attainment lesson or as a lead-in for a new science unit. You can probably think of many other appropriate times when it would be beneficial to focus the students' attention on a problem or concept and to quickly create interest in this problem or concept.

Your role is very simple. You decide on the topic, arrange the groups, start the discussion, and then just observe. After the discussion is over, you might want to discuss with the students ways that the leaders can keep the group focused on the task.

Tutorial. The tutorial discussion group is used most frequently to help students who have experienced difficulties in learning either basic skills or a single scientific concept, or who are absent from science classes. The group has only a few students (usually two to four) and focuses on a narrow range of materials. It is an excellent way to facilitate the handling of manipulatives or to demonstrate and evaluate motor activities.

The arrangements for a tutoring session should provide easy "eyeball to eyeball" contact to ease the flow of communication among all persons. The selected discussion leader is clearly identified and, as such, plays a somewhat dominant role in the group process. This leader has three major functions to perform when in the tutorial mode: (1) questioning the group to pinpoint the exact problem that has blocked learning, (2) providing information or skills to facilitate learning, and (3) encouraging all to ask questions and seek answers among themselves.

Lest you have serious reservations about the tutorial technique, it has been demonstrated that students often learn better from each other than from the teacher! Many school districts currently use student tutors and are finding them to be invaluable resources for the classroom teacher. (Of course, a teacher aide makes an excellent tutor.) I caution, however, that prior to using student tutors, you must be satisfied that each one has mastered the necessary competencies — such as the skills of questioning, giving positive reinforcement, and analyzing work tasks. The student who leads the tutorial science discussion also needs to have developed some skills in the area of human relations. The leader must be patient, yet provide warm and friendly encouragement. The leader must also keep the group moving toward its goal, accept the inputs from those who learn slowly, and prod those group members who are slow to contribute.

Although probably most often used to alleviate student learning difficulties, the tutorial discussion group is an excellent method by which to encourage independent projects for advanced learners. Many gifted students will find it a challenge to try to explain their project to other students.

Task Group. Another easy-to-use discussion type is the *task group*. As the name implies, students are involved in some type of work or activity in which significant contributions can be made by each group member.

Prerequisite to using the task group is the specification of clearly defined tasks or assignments to all group members. The task group is very similar to a committee, having clearly defined goals, individual assignments, and roles. Further, it is beneficial for you to establish a work schedule and a system for internal monitoring of achievements, and initially even to provide all of the learning resources that may be necessary to accomplish the identified tasks. Really, the latter is recommended. At the scheduled conclusion of the task, each subgroup reports its findings to the entire class.

Task groups tend to be teacher dominated in that the teacher usually selects the tasks and assigns each class member to accomplish some specific role. This discussion type can be used very efficiently during the early part of a semester when you are attempting to provide students with specific scientific process competencies. Using this technique you can observe how selected students work with each other and how responsibly they tend to accomplish the task that has been assigned.

In General

The use of small groups in science instruction is one more method by which to make your class more self-sustained. The technique is not to be construed as an "easy way" out of teaching science. After reading the text to this point, you must surely be asking, "Can anyone possibly have the time to do all this and teach science, too?"

The response to that thought is that both science and discussions are chiefly processes. As you teach one process, you integrate the other. My point of view is that teachers should realize that science is easily adapted to any reasonable teaching technique. Further, by developing small-group discussion strategies, you will find that you will have more time to interact with your class members. It will be, of course, an interaction with small groups — but it does increase your interaction significantly.

CHAPTER 5
FITTING THE PIECES ALL TOGETHER FOR SUCCESS

At this point you might just be thinking to yourself, "He hasn't mentioned one thing about needing further science training." That is correct and I had originally not planned to mention it at all. My rationale is that elementary school teachers are expected to be, really, the last of the encyclopedists. Every advocate tends to conclude that you folks need more — more career education, consumer education, drug education, health education, nutrition education, science education, human relations training, counseling training, physical education training, mathematics training, anthropology, economics, history, sociology, and on, and on, and on.

I am an advocate of science education! There will be no apologies for that. Yet, I am a realist. Most of the science activities that you must incorporate into your daily teaching — or more realistically into about two or three hours of instruction per week — require only a general orientation to the basic sciences. To be sure, it is most desirable to have had courses in biology, physical sciences, earth sciences, atmospheric and space sciences, outdoor education, and scientific equipment. Most of you have had one or two courses from that list. Yet, most of you will need some type of added preparation during the in-service phase of your careers.

Organizing Support for Science In-Service Programs[6]

There are four elements that I have found to be essential for success when designing and conducting effective science staff development programs: (1) awareness, (2) application, (3) implementation, and (4) maintenance.

Awareness. Training projects or activities that are designed to provide information about new science concepts, developments, skills, equipment, curriculums, or teaching techniques are classified as *awareness* sessions. These activities are essential to keep all staff members up to date and informed about professionally related skills.

Awareness sessions usually include short conferences, one-day institutes, multimedia presentations, programs given by sales representatives, promotions, lectures, or visitations to schools with a reputation of having a good science program. The information obtained at this level makes one knowledgeable that some science curriculum "exists" or is available for use. The teacher participants are usually learning about selected processes or products for the first time.

In too many cases, in-service programs are centered about one science awareness project after another. Thus, the staff becomes exposed to something new, but that is it — exposed!

While awareness is critical to learning about elementary science programs, an additional set of steps must be taken. Teachers and principals need to put selected innovative science activities into practice — which leads directly to the second element for effective in-service training.

Application. The element of *application* carries the connotation that something is used in a context different from that in which it was initially learned. How does a school district incorporate an application phase into its science in-service programs? There are several alternatives; let us examine but two.

One of the tested methods of application is that of micro-teaching. You select only a few peers or students and try the science lessons with a few of them each day. There is no penalty in micro-teaching for failure as there is in the classroom. Only a few teachers may be involved in the activities. The essential attribute of the application phase is *experience*.

New science curriculums, new science textbook series, and sets of science teaching strategies can be tested by a group of teachers. Perhaps an entire school might act as the trial center for some specific science program, project, or skill-building training. By incorporating the application phase into the school district's in-service operations, a full-scale science in-service program that focuses on a given activity may be supported or aborted, depending on an evaluation of the application phase.

Implementation. Full commitment to a science in-service program takes place after successful application phases. The most intensive period of the project or program is the implementation phase. All *appropriate teachers and administrators are involved* in a full spectrum of in-service activities that relate to the successful teaching and learning of the selected science program. Further, it is during the implementation phase that the supervisory activities of responsible science resource personnel are focused on helping the involved teachers and principals to perfect the science skills, procedures, or content.

The word *appropriate* was carefully inserted in the previous paragraph because it has been often assumed that "everyone" ought to be involved in most school district in-service activity. Such broad scale involvement is seldom, if ever, warranted in any school district. Science in-service programs are planned, designed, and incorporated into the curriculum following carefully conducted needs assessments. It has been my professional experience to observe that needs surveys illustrate a wide variety of desired in-service activities in which specifically designated personnel are the foci. Provide the intensive training only to the immediate users. *Relevance* is the key word in this phase.

One last point: The implementation phase is complete when the involved personnel demonstrate the newly learned and desired science-related behaviors, can perform them in the classroom or office, know why they are being performed, and can evaluate individual proficiency by a set of systematic criteria. The "pay-off" of any systematically conducted science in-service program may be observed through successful individual or group assimilation and implementation.

Maintenance. The final phase is a continued and longitudinal "low-level visibility" set of science in-service activities that follow the major implementation effort. The term *maintenance* implies that some kind of follow-up is required. New persons are hired in the district, persons change grade levels or switch buildings, and some persons will simply need additional training to perfect the needed science skills.

To address these rather special needs, a school district curriculum coordinator must plan and develop a set of internal mechanisms by which to conduct the maintenance in-service activities. One successful method is to identify and reward teachers or administrators who are judged as exemplars of the newly learned science-related behaviors or skills and ask them to conduct short-term clinics or help sessions. A second method is to use the principal or central office resource person (if they have the time) as the training agent. Yet another technique employs multimedia instructional systems. The essential element is to provide all identified individuals with the preparation required for sustained success of the science program.

Implications

School districts that have incorporated the above model tend to find teachers' and administrators' attitudes about science in-service programs to be very positive. The ultimate end, of course, is better science instruction for the student in the classroom.

However, it is abundantly apparent that such science in-service programs require school board commitment to ensure success. This commitment comes first in the form of a *written* board policy that recognizes in-service education as a continuous need of the district. The fiscal commitment of the school board for science will be enthusiastically matched by the professional commitment of teachers and administrators.

This in-service plan implies that learning is truly continuous and is a valued activity. More importantly, it provides a parity for science in-service: The teaching staff and administrative staff each do what is essential for collective success — and for the success of the students, too.

Some Comments About Incentives

Quite obviously, involvement in in-service programs requires more than a passive commitment to the profession. To accomplish any ambitious program requires identification of incentives that may aid professional commitment.

One major intrinsic reward is the self-satisfaction of having become knowledgeable about science. Further, the translation of the scientific knowledge and techniques will help raise student achievement levels.

Other educators will desire to gain college or university credit, ultimately to gain advancement on the salary schedule.

Another incentive is released time. This time is used to observe other teachers, conduct science in-service on the maintenance level, study the adopted science program independently, devise evaluation strategies for the program, or plan for new activities to be incorporated into the science program.

Paying the Bill. School district administrative personnel must be alert for methods to fund in-service efforts. These include linking with a local or regional university to apply to the National Science Foundation for pre-college teacher improvement grants. These grants provide, at no cost to the school district or the participants, science-content–related programs during either the summer or the academic year. The vast majority of these grants provide for instruction to be conducted right in the school district.

State offices of education often have federal projects that can provide some fiscal support for science in-service programs.

The concept of the "Teacher Center" will undoubtedly provide delivery systems to teachers that will improve their science skills, and at the local level.

I would also suggest that a school district contract with sympathetic science educators and scientists at a local or regional university to conduct an in-service effort that is based on clearly specified objectives — determined by the staff. In this manner, the teachers can obtain additional extension credits, the district pays for the training, and the teachers and administrators plan and design the instruction that they need.

There are many other methods of approaching the problem of providing relevant science in-service programs. The essence of in-service is to provide needed experiences to the teachers and principals, to extend and expand their science knowledge, and ultimately to reduce the levels of anxiety that accompany adventures "into the unknown."

Teaching Students with Special Needs

One area of American education that is gaining momentum is the adaptation of instructional materials and science curriculums for handicapped children. These widespread efforts in science education for the handicapped stem primarily from the passage of Public Law 94-142, the Education for All Handicapped Children Act of 1975. Much has been written about PL 94-142, and it must be noted that one aspect of the law has been to make available the totality of the curriculum to all handicapped children.

Public Law 94-142 requires all school districts to prescribe an Individual Education Plan (IEP), detailing the general and specific objectives for the student. Science also requires the specification of objectives. Thus, you already have the element of planning completed.

Assuming that your elementary science program is activity oriented, you will soon find that it is not difficult to accommodate handicapped children in science classes. Recall from the earlier chapters that modern science programs are divergent in characteristic. This same divergence allows for greater flexibility in meeting the special requirements of teaching handicapped children. In many activities there is the need for a small group: two or more children working on the problem; this allows a handicapped child to be placed with a supportive group.

Most activity-oriented elementary school science programs have both physical and biological science components and are designed to be used throughout the year at each grade level, kindergarten through grade 6. A definite teaching cycle is usually suggested. In nearly all modern programs, the initial activities are organized so that the students learn through spontaneous interactions with the various materials. During this stage all students have the opportunity to explore, touch, feel, discuss, and enjoy the stuff of the lesson. Here is the opportunity to initiate those divergent questioning skills, as the students respond divergently when asked what they observe during the initial experiences.

Later, scientific concepts are identified by either you or the students. Again, by having concrete science materials, you will find that the handicapped can attain most concepts as well as your other students.

The last stage of nearly all modern elementary science programs is the application stage. This stage comes only after the concepts have been identified and after the students have had concrete learning experiences that illustrate the respective science concepts. During the final stage, handicapped students will have an equal opportunity to provide examples of where the concept is located in the real world or where it could be applied in a novel situation. With such divergence, there is little chance that either the student or the teacher can fail.

Successful teachers of the handicapped have long known that specificity of tasks is important. You might be worried that a divergent science program will not contribute to the identification of specific learner tasks. To the contrary, even the most divergent programs allow you to prescribe specific learner objectives. Activity-oriented science programs offer a challenge to the handicapped student if manipulative skills are deficient; yet you can adapt the activity so that it may be accomplished by these students.

The essence of science is *doing*. Keep that essence foremost with the handicapped.

I acknowledge that the previous treatment on teaching the handicapped is most underdeveloped. Having handicapped students in the classroom will cause some initial anxieties. Yet, you are not alone. These students have special teachers who will be able to give you explicit methodologies and tips on how to maximize science learning with them.

Evaluating Science Programs

The use of a previously adopted science program or the anticipation of a newly adopted one tends to be a source of elementary teacher frustration and anxiety. If teachers tend to rate the adopted program as being of poor educational quality, then there is a high probability that they will teach that science program as little as possible — if at all. However, one cannot attack any science program without the use of clearly stated and objective evaluative criteria. School districts often have various checklists by which to evaluate science programs. There are, however, far too many cases where adoptions are made somewhat intuitively and the teachers were not truly involved in the adoption. To alleviate these problems, let us now address the issue of science program evaluation.

Program Goals and Objectives. Throughout, I have assumed that present and future elementary school science programs will be activity oriented. Implied in the assumption is that elementary science education will continue to stress the scientific processes by which knowledge is acquired. Thus, a major goal for any program is the development of inquiry as a basis for study.

Whether the developers use performance objectives or some form of knowing what is expected of the learner is critical. While there is great division among science educators about the appropriateness of performance objectives, you need to have some concept of what is to be accomplished with the various activities. This is the essence of stated objectives.

Scope and Sequence. There is the need for some logical development of science activities in a "scope and sequence" chart so that all the staff knows where the program is ultimately going. Traditionally, the concepts of *scope* and *sequence* have meant that a pattern of student learning experiences has been planned or identified. *Scope* implies the relative impact of the subject to be studied. When planning for *depth* or *breadth* of a definite science program, planners determine the scope. There is a great deal of divergence in the scope of topics covered in general science courses being designed for the elementary school; some will treat each topic in depth, while others will address similar topics in a less developed structure. Yet, all topics are selected and designed to provide wider exposure to the students so that *breadth* of the discipline is stressed.

The concept of *sequence* implies that the activities will be structured by a prescribed design. Sequence is usually determined by (1) logic — each topic follows a definite entry and exit; (2) topic — the concepts are sequenced to provide some Gestalt of the subject; (3) hierarchy — each topic is subdivided into specific learning increments, all being arranged in patterns of known to unknown to be expanded from simple to complex; (4) developmental stages — the concepts of developmental psychology, usually Jean Piaget's, are used; and (5) a combination of the four.

All science programs must be examined to determine the scope of topics and the order of sequencing. Since similar topics appear in nearly all books, it is possible to compose a table of topic categories to use in choosing a program that best meets the district's science goals.

Within the concepts of scope and sequence, there should be an analysis to determine if the science program is highly structured. A very highly structured program requires close adherence to the program per se. Such programs allow little in the way of teacher initiative. However, a structured program does have some built-in efficiency — i.e., you know what is expected and can spend your time implementing the program rather than creating new activities. Loosely structured programs, conversely, allow a greater degree of teacher latitude in the teaching of topics.

Accompanying the above is the concept of *program flexibility*. Most programs are built with a K-6 scope and sequence. Other programs tend to have the option of being supplemental to the program that the district has already adopted. In yet other cases, the district can select various components of the science program that ultimately reflect an eclectic approach to science.

Finally, under the concept of scope and sequence comes a determination about the balance of science subjects or topics that are presented. Some programs tend to stress a broad balance between the life and physical sciences. Others tend to stress the physical sciences far more than the life sciences. If a program has a strong life science component, has the district made the necessary commitments to maintain and deliver life science materials to you, or must you maintain your own life science stock? The latter decision is most crucial to the success of any science program that has a life science component.

Instructional Strategies. A good share of this book is devoted to science instructional strategies. However, when evaluating an elementary school science program, you must consider the types of teaching strategies that are required. One important consideration is the relative amount of individualization that will take place. If a program requires a great deal of individual student work, then both you and the students must be taught how to effect that strategy. Because nearly all modern elementary science programs require inquiry in the broadest sense, school district administrative personnel and teachers, alike, must simply plan to conduct in-service efforts that address the many techniques that are included under the generic term of *inquiry*. Programs also vary from those that desire teachers to be very nondirective to those that are highly prescriptive.

Costs and Maintenance. A critical decision area for many science programs concerns the initial cost of the program and the yearly replacement costs for expendable materials. When cost comparisons are made, there is the tendency to weigh the cost of a comparable textbook, which can be kept for about six years on an adoption cycle, against an activity-oriented science program. Ultimately, the evaluation of the alternatives must be made on a *benefit theory* basis since it will cost the district to teach science, no matter what the method. Which program will achieve the intended goals and objectives at the least cost tends to be the basis on which favorable or unfavorable decisions are made.

Teacher Reactions. If your school district has adopted a program or is in the process of field-testing one, then the "using" teachers should be polled to determine their perceptions. This type of evaluation should be designed to measure the interest of the staff, the use of teacher time in teaching science, and the extent to which the teachers perceive that the program is fulfilling the scientific literacy needs of the children. Figure 7 illustrates one instrument that could be used in such an endeavor.

In Final Conclusion

I had as my original goal the reduction of anxiety about teaching science in the elementary schools. I hope that the goal will be met. The real pay-off to you teachers is the actual application of the many ideas and techniques presented herein.

Kindly complete the following survey about our elementary science program.

1. Time survey: Please circle the number of minutes per week spent in the following areas:
   a. Planning time: (0–4)   (5–10)   (11–15)   (16–20)
   b. Setting up materials: (0–4)   (5–10)   (11–15)   (16–20)
   c. Student class time: (0–30)   (31–60) (61–90)   (90–120)
   d. Other science activities: (0–4) (5–10)   (11–15)   (16–20)
   e. The *total* time in minutes per week spent on science _____
2. Is the amount of time *less than*, *equal to*, or *greater than* that spent on mathematics, reading, and social studies respectively? _____

3. Did you find it necessary to supplement the science program with other science materials or activities?
   a. In almost all cases   c. In a few cases
   b. In most cases   d. In no case
4. If supplements were used, specify for which units or lessons: _____

5. What is your overall reaction to the science program?
   a. Very positive   c. Negative
   b. Positive   d. Very negative
6. If negative, please list specifics: _____

7. What unanticipated events have been happening in your science classroom? _____

8. In your opinion how do the students in your classes like the new science program?
   a. Like it very much   d. Dislike it
   b. Like it   e. Dislike it very much
   c. No opinion
9. What concern or problems have you had with the program? _____

Figure 7. Sample Instrument To Evaluate Science Programs

To be certain, I could have presented other topics that would reduce teacher anxiety. However, this work must be viewed as a "commencement" — a beginning to make you aware of the potentials that exist in your classroom and to stimulate you to analyze your own specific needs, wants, desires, and anxieties. Many of the detailed accounts on the previous pages should lead you to examine the current assets and deficits of your school's elementary science program, as well as the assets and deficits of your own professional background.

In the final analysis, it will be the teachers and principals who will improve science in our elementary schools. No curriculum is "teacher-proof." The single most important person in the classroom (other than the students, of course) is the teacher. Teachers inspire. Teachers provide the affective dimensions of science. Teachers challenge, guide, and share in the richness and joy of young learners accomplishing their science lessons and making discoveries that are for them truly significant events. Yes, it is the human side of science teaching that is all important. Inspiration for excellence begins in the kindergarten and continues in a never-ending cycle. Through the field of science you have an advantage over all other teachers or subjects — science is the shape of things to come!

Yet, the real anxiety of teaching science lies in our own inability to live peacefully on a planet that will shortly be inhabited by six billion people. Humanistic concepts such as charity, love, and justice need operational definitions just as gravity, fusion, and relativity. A critical challenge in the teaching of science is to reduce irrational patterns in human behavior through the processes of scientific thought. How to discover and apply new knowledge and extend it to benefit all humankind is the challenge that great teachers have always presented to their students. Let that be, in some small part, your challenge!

REFERENCES

[1] The author acknowledges support from the Pre-College Teacher Development in Science Program of the National Science Foundation, Grant #SPI 78-03954, in the preparation of this paper. The work of Dr. James Migaki is also observable in the next series of ideas. The author gratefully acknowledges his generous contribution of techniques.

[2] Benjamin S. Bloom, and others, *Taxonomy of Educational Objectives: Cognitive and Affective Domains* (New York: David MacKay Co., Inc., 1969).

[3] Mary Budd Rowe, *Teaching Science as Continuous Inquiry* (New York: McGraw-Hill, 1973).

[4] Jacob S. Kounin, *Discipline and Group Management in Classrooms* (New York: Holt, Rinehart, and Winston, 1970).

[5] J. Richard Suchman, *Inquiry Development Program in Physical Science* (Chicago: Science Research Associates, 1966).

[6] The major theses presented in this section are adapted from Donald C. Orlich, "Conducting Effective In-Service Programs by Taking . . . 'AAIM,'" *The Clearing House* 53 (September 1979).

RECOMMENDED READINGS

Cornbleth, Catherine. "Student Questioning as a Learning Strategy." *Educational Leadership* 33, no. 3 (December 1975): 219–222.

Davis, O.L., Jr., and Hunkins, F. P. "Textbook Questions: What Thinking Processes Do They Foster?" *Peabody Journal of Education* 43 (March 1966): 285–292.

Kounin, Jacob S. *Discipline and Group Management in Classrooms.* New York: Holt, Rinehart, and Winston, 1970.

Orlich, Donald C., and Magaki, James M. "Guided Inquiry, Questioning and Elementary Science Teaching." *Washington Science Teachers' Journal* 19, no. 2 (Spring 1979): 9–13.

Orlich, Donald C., et al. "A Change Agent Strategy: Preliminary Report." *The Elementary School Journal* 72, no. 6 (March 1972): 281–293.

Rist, Ray C. "Student Social Class and Teacher Expectations: The Self-Fulfilling Prophecy in Ghetto Education." *Harvard Educational Review* 40 (August 1970): 411–451.

Rowe, Mary Budd. *Teaching Science as Continuous Inquiry.* New York: McGraw-Hill, 1973.

Suchman, J. Richard. *Inquiry Development Program in Physical Science.* Chicago: Science Research Associates, 1966.

Verduin, John R., ed. *Conceptual Models in Teacher Education.* Chapter 10, "Structure of the Intellect," citing the work of James Gallagher and associates (p. 93). Washington, D.C.: American Association of Colleges of Teacher Education, 1967.